W9-AEU-949

INTERDISCIPLINARY MATHEMATICS

VOLUME XI

GEOMETRIC STRUCTURE OF SYSTEMS-
CONTROL THEORY AND PHYSICS
PART B

ROBERT HERMANN

MATH SCI PRESS
53 JORDAN ROAD
BROOKLINE, MASSACHUSETTS

Copyright © 1976 by Robert Hermann
All rights reserved

Library of Congress Catalog Card No. 74-30856
ISBN   0-915692-14-7

MATH SCI PRESS
53 JORDAN RAOD
BROOKLINE, MASSACHUSETTS

Printed in the United States of America

MATH.-SCI.

QA
402
.H43
Pt. B

1594759

PREFACE

I carry on with my program of understanding the geometric foundations of engineering and physics. Chapters 1-6 concern optimal control theory, 7-12 mechanics, and 13-15 the geometry of "optimization", a subject which is of increasing importance throughout engineering and economics. I usually use the Prefaces to my books for my Editorials. For this volume, I have put them into a separate essay which follows. Again, I refer to my previous books by abbreviations, which are listed in the bibliography. Some of the thought that went into this work was supported by the Ames Research Center (NASA) and the National Science Foundation Grant No. MCS75-07993. I particularly thank Brian Doolin of NASA. I thank Sheldon Glashow and John Stachel for their hospitality in the physics departments at Harvard and Boston Universities, and Karin Young for her superb typing.

In this volume, I have added a new feature--Final Remarks. (I was tempted to call it an Anti-Preface, but resisted.) As I finish one book in this series, I find my thought evolving, often developing into a treatment of related material in a later book. My justification for inflicting this on the reader is that often many different sorts of insights are needed for the sort of complicated structure I am trying to build.

Contents

An Essay on Interdisciplinary Mathematics

by

R. Hermann

This is the twenty-first book I have written, and the tenth anniversary of my first book, "Lie Groups for Physicists". To celebrate, I will attempt to describe some general features of my program and its future directions.

I mean IM (the abbreviation for "Interdisciplinary Mathematics") to deal specifically with the interactions between contemporary pure mathematics and those areas of physical science and engineering (and, possibly, biology, economics, ...) that use mathematical ideas in a significant way. For most creative pure mathematicians, the application of their work is left to some vaguely defined future. It is almost considered to be a random process! In fact, I believe that the process can be nurtured more systematically, and that mathematics itself will benefit from the stimulus of problems that are not simply generated within the closed world of the pure mathematician. There has been some movement towards opening up this world to outside stimulii, but I cannot see that anything substantial has changed. If anything, at a practical level it has become worse, since it is not so much harder for a young, unestablished pure mathematician to find the freedom to become interested in applications! I can testify to how much time it takes. In my day one's "punishment" for trying to do something of this sort was to be denied a job in a mathematics department at a leading university (presumably, this is why mathematics oriented towards applications has been stultified in American academia), but today one will probably not find a job, period!

How does IM differ from "Applied Mathematics"? Now, there are (at least) two different sorts of "applied mathematics". The first kind is concerned with the mathematics that is needed in the short term to deal with an area of application. For example, this might involve a branch of mathematics that has been by-passed by the mainstream of pure mathematicians. A common situation today is that our greatly increased computational ability suggests the revival and renewal of approaches to classical problems that were formerly considered to be too difficult.

The seond sort involves a much greater degree of
emphasis on the interaction between mathematics and the
conceptual and structural problems encountered in a
discipline. Ideally, the phenomena under investigation
suggests new mathematical ideas, and these ideas play a
creative role in understanding and describing the
underlying concepts. A good deal of mathematical physics
is of this second type. (Unfortunately, a creative interaction between mathematicians and physicists is often
discouraged by various cultural and sociological factors.)
A less traditional example is provided by mathematical
systems and control theory. Biology is apparently not
yet in this stage, but everyone believes it will arrive
there some day. The possibility of this sort of interaction between mathematics and the social sciences is
more controversial (some believe that it is dubious),
but there are already certain relatively small points
of contact that might be expanded in the future.
IM is closely allied to this second type of applied
mathematics, but is closer to pure mathematics itself.
I see it as directly using the intellectual techniques
developed in pure mathematics in the task of providing
mathematical structure to engineering, physics, biology,...
("Applied Mathematics", of either variety, usually has its
own intellectual structure, which only incidentally
impinges on that of the pure mathematician.) Implicit
within this identification with pure mathematics in terms
of its intellectual techniques is the hope that the
Interdisciplinary Mathematician will be well-trained in
pure mathematics itself--indeed, to a much greater
breadth than most pure mathematicians--and perhaps even
use his insights in fields of applications to develop
new or improved mathematical tools.
Here are certain historical examples. The first is
the development of tensor analysis differential geometry
in the 1920's, obviously related to the drama and excitement of Einstein's work on relativity. The best mathematical work in this period was by Cartan and Weyl. They
utilizied the excitement stirred up by physics to develop
extremely important ideas in differential geometry and
Lie group theory, and conversely their work suggested new
mathematical approaches to physics. (Of course, physicists paid very little attention at the time!)
Curiously, quantum mechanics had a much less direct
effect on mathematical creativity. Of course, there were
the great books by von Neumann and Weyl, but that was more
or less the end as far as professional mathematics went,
despite the obvious fact that quantum mechanics is by far

the most mathematically sophisticated area of applied mathematics! Only recently has there been a return to this interaction, which should have been pursued in the 1930's, 40's and 50's--physicists have moved on to other problems.

Another striking example is Wiener's interest in the application of Fourier analysis and probability theory to electrical engineering and communication theory, which has now expanded to awesome proportions. His work was brilliantly carried on by R. Kalman, and the "Kalman filter" has probably been the most successful single example in modern times of the application of modern mathematics. The RIAS, Berkeley, and Bellman schools (building, of course, on the intitial work of the Pontrjagin group) had a great intellectual success, particularly in persuading engineers that "state space" methods were a superior way of thinking about optimal control problems, especially in terms of design, and utilization of computers.

To a certain extent, I have been influenced by the example of Bourbaki; for example, I hope that the very existence of material in print will have a long range effect. It is probably impossible to write the "Bourbaki of Applied Mathematics" (although it would be fun to try!), but I think that efforts such as mine can be of help. I have begun two institutions -- Math Sci Press and the Institute for Mathematical Engineering and Physics-- that I hope will carry out some of this work. Of course, I fundamentally disagree with Bourbaki's approach to mathematics, since it is based on rejection of the very sort of outside interaction and stimulating mathematical activity that I want to encourage! I hope that the world of mathematics can synthesize the modern emphasis on technical sophistication with classical concern for breadth, intuition and attention to the problems of the outside world.

Comments on Mathematical Physics and Engineering

by Robert Hermann

Much of this series of books is my personal attempt to bring the full force of modern geometry to bear on a wider spectrum of problems in physics and engineering. While engaged in this task, I have developed rather strong feelings (mainly negative) about the various forces in the scientific world which impinge on this work; I will take the opportunity to express them, in the hope that things might improve. These thoughts may seem angry and aggressive, but let me assure the reader that this is only "philosophical" anger and aggression!

Begin by contemplating the present state of the relations between mathematics and physics. The most striking fact is how the traditional interconnection has degenerated! It is evident that, of all the fields, physics and mathematics have the stronest historical affinity for each other. Most of the important problems of mathematics have arisen from physics--indeed, as von Neumann and Weyl themselves emphasized, this flux of problems from physics to mathematics is probably the only antidote to the degeneration of mathematics into an intellectual game, like chess. Most people who are at all exposed in their youth to training on the frontiers between mathematics and physics--and this includes many mathematicians, biologists, chemists, economists, etc. who never go on into mathematical physics--come away with a profound feeling that they really understand how Nature can suggest the right mathematics. (For this reason, if no other, it is a tragedy that mathematicians now receive no training in physics!) Unfortunately, the reasons for the degeneration in this connection can probably only be adequately understood in terms of a full scale historical study. Obviously, there was much arrogance on both sides, and both disciplines are paying for it today--the mathematicians in their obsession with sterile formalism (the solving of Great Problems arising within mathematics is no substitute!), the physicists in their inability to make progress in fundamental physics which is in any way comparable to the great accomplishments of the 19-th or the first quarter of the 20-th centuries. Now, there has been a modest revival in terms of groups of people who call themselves Mathematical Physicists, but they often have narrow obsessions of their own in terms of functional analysis which seems to prevent them from serving as the bridge between mathematics and physics. It is curious to me that they do so little work in relation to modern elementary particle physics,

but concentrate on mathematical problems of the physics that
was essentially complete in 1930. Of course, there is their
grand goal of mathematically understanding Quantum Field
Theory--either in terms of "axiomatization" or "construction".
As presently constituted, I do not see that either branch
is that interesting to justify the labor that is going
into it. It is as if Newton, after writing down F = MA,
spent the rest of his life trying to prove the existence
theorem for solutions of ordinary differential equations,
instead of finding examples which he could understand
mathematically far enough to check with Nature. Of course,
the theoretical physicist who works in Quantum Field
Theory--Elementary Particle Physics seems to be equally in
a rut--(and it is shocking that most do not worry at all
about the intellectual and mathematical integrity of
their discipline). He can do no better than try to understand non-linear equations by means of perturbation series
(which usually are meaningless anyway) in terms of linear
equations. Again, the comparison with Newton or Einstein
(whose great accomplishments were precisely to understand
the non-trivial mathematical basis of their Equations far
enough to derive meaningful physical conclusions) makes the
best work being done in fundamental physics today seem totally
inadequate to what will obviously be required to solve the
Elementary Particle riddle.

Of course, I realize that in criticizing an "obsession"
with Functional Analysis among Mathematical Physicists, I am
leaving myself open to the counter charge that I am "obsessed"
with Differential Geometry. The cynical reader will no
doubt think of the Blind Men Trying to Describe the Elephant.
However, I do feel that this branch of mathematics (in
alliance with certain branches of algebra, and with certain
more "concrete" parts of functional analysis) has the most
to offer to applications across a broad spectrum. Again,
because of complicated historical and academic-political
factors, which I feel, but cannot describe more precisely,
the reality has been quite perverse--as a branch of mathematics, geometry has had the least to do with applications,
and physicists, engineers, biologists, economists,... face
an insuperable barrier to learning the basic ideas. Obviously,
all of my books are basically devoted to the amelioration
of this unfortunate situation. I fully expect that it will
take many years before their ideas come to fruition!

I do feel very positive towards a certain type of
functional analysis--what Paul Levy called "concrete" functional analysis. Here, one uses functional analysis less
as a rigorous mathematical tool, but more as an intellectual
device for unifying diverse topics, extending arguments to
infinite dimensional situations, etc. Of course, I tend to

cherish those parts of functional analysis that fit in well with geometric situations and reasoning--and I suspect that this is the part that is most useful, and that fits in best with the intuition with which a physicist or engineer approaches a problem. Because mathematicians have insisted on making functional analysis such a formidable and "rigorous" obstacle course, many engineers and physicists do not even feel comfortable with the most elementary level of functional analysis methodology--say, thinking in a coordinate-free, basis-independent way.

One of the most discouraging things about trying to teach applications of modern mathematics to non-mathematicians is their lack of training in this direction. This is particularly so for physicists--engineers actually have improved tremendously in this ability to get beyond stage zero in the use of modern mathematics. Here we also encounter the interesting difference between Engineering and Physics and (because science is so gloriously personal, despite the myth otherwise) between engineers and physicists. Again, to go into these differences--and their implications for mathematics--would involve a full-scale work. I can only say that I respect more some of the work of the last twenty years in what one might call Mathematical Engineering than that in Mathematical Physics. The Hare and the Tortoise comes to mind--physicists are usually much more intellectually impressive than engineers, but the tortoise seems to have won again. Perhaps a basic reason is that many engineers receive a far better graduate education in modern mathematical ideas! (For example, the reader will notice that I often recommend books by engineers-- and often even textbooks--as the best treatment of certain mathematical subjects, on the right intermediate level of abstraction, and mathematical sophistication tempered with concreteness and a sense of reality. This is in utter contrast with physics, where there is nothing comparable, i.e., there are no books on classical or quantum mechanics, electromagnetism, elementary particles, etc. that are even remotely adequate for someone trying to understand Nature with the tools of modern mathematics.) Of course, there are other factors in the differences between engineering and physics. (For example, the nature of the disciplines; precisely because engineers are much less constrained than physicists--they must build something that works adequately, not figure out a fantastically complicated puzzle presented by Nature--they can use mathematical schemes more freely. Also, the advance in computer technology has enabled them to implement mathematical ideas that would have been laughed away by previous generations of engineers as hopelessly impractical!)

Unfortunately, most of the progress that seems to have been made in the 1960's in developing an impressive body of Mathematical Engineering seems to be slipping away, just as it was beginning to have influence and scientific success. The call for "relevance" in the early 1970's has had the usual perverse effect of stimulating short-term "pragmatism", while discouraging the sort of long-term theoretical work that is "relevant" in the best <u>scientific</u> sense. Presumably, this is a corollary of one of Parkinson's Laws.

Now, I must say that one of the reasons for the relative stagnation of this high-level sort of Mathematical Engineering seems to be the indifference of the pure mathematicians. They still have <u>some</u> appreciation of the traditional relation between mathematics and physics--what they lack is knowledge and insight!--but have an attitude that one must almost characterize as snobbish towards Engineering--say the attitude of the second son of a Lord who has to think about going into Commerce. (This was almost literally true--many of today's leading Pure Mathematicians did Applied work in World War II as an alternative to the Army.) Such attitudes are, of course at the root of many of the difficulties in Applied Mathematics taking root in the American academic soil--and we all know who is going to pay for it in unemployment. Equally serious, on the intellectual level, professional mathematics is deprived of many interesting and fascinating problems aris- ing from the rapid growth of technology.

Despite all of this gloom, I want to end on a positive note. Mathematics does progress today--as even I must admit. In one sense, it is much better to be a working mathematician today that ever before--even than the Heroic Age of Geometry, say 1880-1900--we understand so much more, in what must be a much less painful way. I do this screaming because I do not see even the minimal steps being taken towards putting all of this to work--but I suppose it must eventually happen.

Chapter 1

THE CARATHEODORY APPROACH TO OPTIMAL
CONTROL VARIATIONAL PROBLEMS,
AND THE PONTRJAGIN MINIMAL PRINCIPLE

1.   INTRODUCTION

The Caratheodory approach to the calculus of variations
is usually developed in the context of the "Lagrange variational problem". (See his book [1], and DGCV.) In Volume III,
Chapter 6, I have briefly indicated how the ideas could be
adapted to the standard control-theoretic variational problems.
In this chapter I shall redo some of his work in a more
systematic way. The way the theory is developed here (it
has already been more briefly presented in Volumes III and IX)
is meant to lead into the applications in mathematical economics.
(See Intriligator [1].) This will require treatment of the
transversality (i.e., boundary condition) properties, with
particular attention to the "infinite-time horizon" problems
that appear in economics. In fact, I believe that this problem
is a prime candidate for the application of these geometric
ideas!

I will now describe a standard set of notations which
will be used throughout this chapter. X is a (finite

1

dimensional) real vector space, called the <u>state space</u>. U
is another vector space, called the <u>control space</u> or the
<u>input space</u>. t is a real variable, identified with "time",

$$-\infty < t < \infty.$$

The space of all such t will be denoted by

T.

T may be called the <u>time space</u>.

<u>Remark</u>: To handle other situations, e.g. describe time
intervals, it is often convenient to allow T to be a more
general set then the real numbers.

A <u>control dynamical system</u> is defined by a mapping

$$\varphi: X \times U \times T \to X \qquad\qquad (1.1)$$

A curve $t \to (x(t), u(t)$, in $X \times U$ is a <u>trajectory</u> of this
dynamical system if it satisfies the following set of
ordinary differential equations:

$$\frac{dx}{dt} = \varphi(x(t), u(t), t). \qquad\qquad (1.2)$$

A <u>Lagrangian</u> is a map

$$L: X \times U \times T \to R. \qquad\qquad (1.3)$$

A <u>performance criterion</u> for curves $t \to (x(t), u(t))$,
$t_0 \le t \le t_1$, in $X \times U$ which are solutions of (1.2) is a

real-valued function of the form:

$$\int_{t_0}^{t_1} L(x(t), u(t), t)dt + f(x(t_1), t_1), \qquad (1.4)$$

where L is a Lagrangian, and f is a map,

$$f: X \times T \to R.$$

If L is a Lagrangian, and

$$t \to (x(t), u(t)), \; t_0 \le t \le t_1,$$

is a curve in $X \times U$, the real number

$$\int_{t_0}^{t_1} L(x(t), u(t), t)dt \qquad (1.5)$$

is called the <u>action</u> of L along the curve.

We shall often use coordinate systems for X and U. Denote them by:

$$x = (x_i), \; 1 \le i, j \le n = \dim X \qquad (1.6)$$

$$u = (u_a), \; 1 \le a, b \le m = \dim U. \qquad (1.7)$$

<u>Remark</u>: Ultimately, one will want to replace this "local" situation by a "global" one. Here is a sketch of one way to do this.

Replace X by a manifold, even in a rather "generalized" sense. (For example, an "infinite dimensional manifold," such as a Hilbert space or a manifold modelled on a Hilbert

space, or a "finite dimensional manifold" with "corners"
and "boundaries.") Replace X × U × T by a fiber space

$$\pi: E \to X \times T.$$

"Lagrangians" L are then real-valued functions on E. "Control systems" are then fiber preserving maps

$$\varphi: E \to T(X),$$

where T(X) denotes the <u>tangent vector bundle</u> to X.

2.   COMPLETELY INTEGRABLE LAGRANGIANS

Keep the notations introduced in Section 1. Among
the Lagrangian functions, there is a special class - the
"integrable" ones - which play a special role in the
Caratheodory theory. They are essentially determined by
the condition that the action along a curve X × U only
depends on the end-points of the projected curve in X.

<u>Definition</u>. A Lagrangian L: X × U × T → R is said to be
<u>completely integrable</u>, (with respect to the control system
(1.2)) if there is a function

$$S: X \times T \to R$$

such that:

$$\int_{t_0}^{t_1} L(x(t),\ u(t),\ t)dt = S(x(t_1),\ t_1)$$
$$- S(x(t_0),\ t_0) \tag{2.1}$$

for each curve $t \to (x(t),\ u(t))$ in $X \times U$ which satisfies (1.2), i.e. which is a trajectory of the control system. The function S is called the <u>action function</u> associated with the Lagrangian L.

Such integrable Lagrangians exist, whatever the control law. Hence is one method for constructing them:

<u>Theorem 2.1.</u> For <u>any</u> function

$$S:\ X \times T \to R,\ \text{let}\ L:\ X \times U \times T \to R$$

be defined in the local coordinates (1.6), (1.7) by the following formula:

$$L(x,\ u,\ t) = \frac{\partial S}{\partial t} + \frac{\partial S}{\partial x_i}\ \varphi_i(x,\ u,\ t), \tag{2.2}$$

where φ_i are the functions which are the components of the map φ occurring in the differential equations (1.2). Then, L is completely integrable (with respect to the control system (1.2), with S as action function.

<u>Proof.</u> With L defined by (2.2), we must show that (2.1) is satisfied. Let

$$t \rightarrow (x(t),\ u(t),\ t),$$

$$t_0 \le t \le t_1,$$

be a solution of the control differential equations (1.2).
Then,

$$S(x(t_1),\ t_1) - S(x(t_0),\ t_0)$$

$$= \int_{t_0}^{t_1} \frac{d}{dt}\ S(x(t),\ t)dt$$

$$= \int_{t_0}^{t_1} (\frac{\partial S}{\partial t} + \frac{\partial S}{\partial x_i}\ \frac{dx_i}{dt})dt$$

$$= \int_{t_0}^{t_1} (\frac{\partial S}{\partial t} + \frac{\partial S}{\partial x_i}\ \varphi_i(x(t),\ u(t),\ t))dt.$$

$$= \int_{t_0}^{t_1} L(x(t),\ u(t),\ t)dt,$$

where L is defined by (2.2).

q.e.d.

Remark: In case the system (1.2) is what I earlier called
a Caratheodory-Chow controllable system, I can prove that
conversely any completely integrable Lagrangian must be of
the form (2.2). Since this result does not play any
immediate role in our work, I shall put it off to a later
volume.

3. EXTREMAL FIELDS AND HAMILTON-JACOBI-CARATHEODORY FUNCTIONS

Let L, L': $X \times U \times T \to R$ be Lagrangians. Suppose a control system (1.2) is given.

Definition. L is said to <u>dominate</u> L' if the following inequality is satisfied:

$$L(x, u, t) \geq L'(x, u, t) \tag{3.1}$$

for all $(x, u, t) \in X \times U \times T$.

Of course, if inequality (3.1) is satisfied, it can be integrated, to give a similiar inequality for the action function on curves, defined by the Lagrangians L, L':

$$\int_{t_0}^{t_1} L(x(t), u(t), t)dt$$

$$\geq \int_{t_0}^{t_1} L'(x(t), u(t), t)dt. \tag{3.2}$$

Thus, (3.2) is a trivial consequence of (3.1). However, it is a key point in Caratheodory's treatment of the calculus of variations.

Definition. A <u>control law</u> is a mapping

$$\gamma: X \times T \to U. \tag{3.3}$$

A curve $t \to x(t)$ in X is a <u>trajectory of the control law γ</u>

(relative to the control system (1.2)) if the following
differential equations are satisfied:

$$\frac{dx}{dt} = \varphi(x(t), \gamma(x(t), t), t). \qquad (3.4)$$

Remark: Suppose

$$t \rightarrow x(t)$$

is a curve in X which is a solution of (3.4). Set:

$$u(t) = \gamma(x(t), t).$$

Then, equation (3.4) implies that:

t → (x(t), u(t)) is a control system
trajectory, i.e. a solution of equations (1.2).

Definition. Let L: X × U × T → R be a Lagrangian. A
triple (L', S, γ) consisting of an integrable Lagrangian
L', with action function S, and control law γ, is said to
be an extremal field for L if the following conditions are
satisfied:

L dominates L', i.e.
                                                              (3.5)
(3.2) is satisfied.

$$L(x, \gamma(x, t), t) = L'(x, \gamma(x, t), t). \qquad (3.6)$$

If these conditions are satisfied, then γ is called an

extremal control law, and S is called a Hamilton-Jacobi-
Caratheodory function.

Historical Remark: I believe the historical roots of what
I am calling and expounding as "Caratheodory's method" lie
in Hilbert's work on the calculus of variations. R. Bellman
extended Caratheodory's work to cover more general situations, e.g. inequality constraints, discrete time intervals,
etc. He called this dynamical programming. Thus, if one
wanted to cover all these historical bases, S might be
called a Hamilton-Jacobi-Hilbert-Caratheodory-Bellman
function. However, I will not burden the reader with this
terminology. In fact, I will usually leave Caratheodory's
name off also, and abbreviate to:

            H-J function.

    Here are the main results concerning these concepts.

Theorem 3.1. Let (L', S, γ) be an extremal field (relative
to the control system (1.2) for the Lagrangian L. Then,
for any curve $t \rightarrow (x(t), u(t))$, $t_0 \leq t \leq t_1$, satisfying
(1.2), the following inequality holds:

$$\int_{t_0}^{t_1} L(x(t), u(t), t)dt \geq S(x(t_1), t_1) \\ - S(x(t_0), t_0)$$

(3.7)

Further, if $t \rightarrow x(t)$ is a trajectory of the control law γ, i.e. if (3.4) is satisfied, then equality holds in (3.7).

    Proof. Start with inequality (3.2). Insert formula (2.1) on the right hand side of (3.2). This gives (3.7), (3.6), the defining property of "extremal field," when combined with (2.1) and (3.4), gives the equality sign in (3.2).

    Notice that an extremal field gives us a formula of curves $t \rightarrow x(t)$ in X which are "extremal curves" for the variational problem, in the sense that this is a control function $t \rightarrow u(t)$ such that

$$t \rightarrow (x(t), u(t))$$

gives a minimum to the action function defined by the Lagrangian L. Let us use this to make a formal definition of "extremal."

Definition. A curve $t \rightarrow (x(t), u(t))$ $t_0 \leq t \leq t_1$, in X × U is an extremal curve (in the sense of Caratheodory) of the variational problem defined by the Lagrangian L and the control system (1.2) if there is an extremal field (L', S, γ) such that the curve $t \rightarrow x(t)$ is a trajectory of the control law γ, such that:

$$u(t) = \gamma(x(t), t)$$

$$\text{for } t_0 \leq t \leq t_1. \qquad (3.8)$$

Here is a main property of "extremals."

Theorem 3.2. Let $t \rightarrow x(t)$, $u(t)$, $t_0 \leq t \leq t_1$, be an extremal curve. Let $t \rightarrow (\hat{x}(t), \hat{u}(t))$, $t_0 \leq t \leq t_1$, be any other curve in $X \times U$ such that:

$$\hat{x}(t_0) = x(t_0)$$

$$\hat{x}(t_1) = x(t_1). \qquad (3.9)$$

Then,

$$\int_{t_0}^{t_1} L(x(t), u(t), t)dt$$

$$\leq \int_{t_0}^{t_1} L(\hat{x}(t), \hat{u}(t), t)dt. \qquad (3.10)$$

In particular, $\int_{t_0}^{t_1} L(x(t), u(t), t)$ is the <u>minimum value</u> taken on by the action, among all curves in $X \times U$, parameterized by $t_0 \leq t \leq t_1$, whose projection in X starts and ends at the same point. In other words,

$$t \rightarrow (x(t), u(t))$$

is the solution of the problem of <u>minimizing the action</u> <u>among curves with fixed end-points.</u>

Proof. The last remarks follow from (3.10). To prove
(3.10), apply inequality (3.7) to the curve $t \to (\hat{x}(t), \hat{u}(t))$,
assuming that (3.9) is satisfied:

$$\int_{t_0}^{t_1} L(x(t), u(t), t)dt$$

$$\geq S(x(t_1), t_1) - S(x(t_0), t_0)$$

$$= S(x(t_1), t_1) - S(x(t_0), t_0)$$

(because of (3.9)), =, the fact that equality now holds on
(3.7) since $t \to x(t)$ is a trajectory of the control law,

$$\int_{t_0}^{t_1} L(x(t), u(t), t)dt.$$

Remark: As a bonus from this argument, we see the geometric
meaning of the H-J function S. It can be described as
follows:

    For fixed $(x_0, t_0) \in X \times T$,
    $S(x_1, t_1)$ is the value of the action
    along the extremal curve which starts
    at (x_0, t_0) and ends at (x_1, t_1).

This property is what Bellman calls the "optimality property." It can be used directly in certain situations to
obtain an intuitive feeling for the situation, but I be-

lieve it is better to couple it with a more formal description of the underling geometry, as we have been doing.

4. THE PARTIAL DIFFERENTIAL EQUATION FOR THE HAMILTON-JACOBI-CARATHEODORY FUNCTION

Continue with the notations of Section 3. Suppose that X and U are manifolds, with coordinates:

$$x = (x_i), \quad 1 \leq i, \; j \leq n = \dim X$$

$$u = (u_a), \quad 1 \leq a, \; b \leq m = \dim U.$$

Suppose the control system

$$\varphi: X \times U \times T \to T(X)$$

is given by the following functions:

$$\dot{x}_i = \varphi_i(x, u, t). \tag{4.1}$$

Let

$$L: X \times U \times T \to R \tag{4.2}$$

be a Lagrangian function. Let

$$S: X \times T \to R$$

be another real-valued function. Set:

$$L' = \frac{\partial S}{\partial t} + \frac{\partial S}{\partial x_i} \, \varphi_i \tag{4.3}$$

Then, L' is another Lagrangian. We have seen that L' is
completely integrable, with S as action function.

Recall that S is a <u>Hamilton-Jacobi-Caratheodory func-</u>
<u>tion for L</u> if the following two conditions are satisfied:

$$L \geq L'. \tag{4.4}$$

There is a control law map γ: X × T → U such that:

$$L'(x, \gamma(x, t), t) = L(x, \gamma(x, t), t)$$
$$\text{for all } (x, t) \in X \times T. \tag{4.5}$$

To find the differential equations, observe that (4.4)
and (4.5) <u>together</u> imply that:

For fixed (x, t) $\in$ X × T, the point
γ(x, t) $\in$ U is an absolute minimum
point of the function   (4.6)
u → L(x, u, t) - L'(x, u, t).

To exploit (4.6) in terms of differential equations, we
<u>must assume that U is a smooth manifold.</u>

<u>Remark</u>: In many applications, U will be a more general
space, e.g. a manifold with boundary, with the minimum
occurring on the boundary. This leads to what the engineers
call <u>bang-bang control</u>. In this situation, Bellman's theory
of <u>dynamic programming</u> takes over, but it is not as

tractable analytically.

The first order conditions for (4.6) are then:

$$\frac{\partial}{\partial u_a} (L-L')(x, \ \gamma(x, \ t), \ t) = 0 \qquad (4.7)$$

Use the formula (4.3) for L':

$$\frac{\partial L'}{\partial u_a} = \frac{\partial S}{\partial x_i} \frac{\partial \varphi_i}{\partial u_a}. \qquad (4.8)$$

Hence, condition (4.7) becomes:

$$0 = \frac{\partial L}{\partial u_a} (x, \ \gamma(x, \ t), \ t)$$

$$- \frac{\partial S}{\partial x_i} (x, \ t) \frac{\partial \varphi_i}{\partial u_a} (x, \ \gamma(x, \ t), \ t). \qquad (4.9)$$

Condition (4.5) becomes:

$$\frac{\partial S}{\partial t} + \frac{\partial S}{\partial x_i} \varphi_i(x, \ \gamma(x, \ t), \ t)$$

$$= L(x, \ \gamma(x, \ t), \ t). \qquad (4.10)$$

Conditions (4.9)-(4.10) constitute a system of coupled differential-implicit equations for the functions $(S(x, \ t), \ \gamma_a(x, \ t))$. They may be called the Hamilton-Jacobi-Caratheodory equations. We shall not deal with them directly, but shall convert them into the traditional (in physics) "Hamilton-Jacobi equation," by introducing a Hamiltonian formula.

5. THE COSTATE VARIABLES, THE HAMILTONIAN FUNCTIONS, AND
    THE CLASSICAL HAMILTON-JACOBI EQUATION

Continue with the notations of Section 4. In addition
to the variables (x_i, u_a, t), introduce the variables (y_i),
called the costate variables, parameterizing a space Y.
Set:

$$H'(x, u, y, t) = L + y_i \varphi_i. \tag{5.1}$$

Let M' be the following space:

$$M' = X \times U \times Y \times T. \tag{5.2}$$

H' is a real valued function on M', called the Hamiltonian.

Remark: This particular way of constructing a "Hamiltonian"
is due to Pontrjagin, in connection with his famous
"Maximal Principle," (which in our way of doing it will
appear as a "minimal principle.") Consider the following
functions on M':

$$\frac{\partial H'}{\partial u_a} \equiv \frac{\partial L}{\partial u_a} - y_i \frac{\partial \varphi_i}{\partial u_a}. \tag{5.3}$$

Definition. The variational problem defined by the Lagrangia
L and control system φ is said to be regular if the following condition is satisfied:

$$\det\left(\frac{\partial^2 H'}{\partial u_a \partial u_b}\right) \neq 0. \tag{5.4}$$

Theorem 5.1. If the variational problem is regular, then
the functions

$$\frac{\partial H'}{\partial u_a}$$

on M' are functionally independent, i.e. their differentials
are everywhere independent 1-forms.

   Proof.

$$d\left(\frac{\partial H'}{\partial u_a}\right) = \frac{\partial^2 H'}{\partial u_a \partial u_b}\, du_b + \ldots \tag{5.5}$$

(The terms ... in (5.5) denote terms in

$$dx_i, \quad dy_i, \quad dt).$$

If the $\frac{\partial H'}{\partial u_a}$ were not regular, there would be (at each point
of M') real numbers λ_a such that:

$$\lambda_a d\left(\frac{\partial H'}{\partial u_a}\right) = 0.$$

Combining this with (5.5), we see that:

$$\lambda_a \frac{\partial^2 H'}{\partial u_a \partial u_b} = 0,$$

Contradicting (5.4).

   Now, set:

$$M = \text{set of points of M' on}$$

$$\text{which the } \frac{\partial H'}{\partial u_a} \text{ are zero.} \tag{5.6}$$

Theorem 5.2.   If the variational problem is regular, then M is either the empty set, or is a regularly embedded submanifold of M'.

Proof.   This follows from the Implicit Function Theorem.

Remark: Recall that "M is a regularly embedded submanifold" means that M with the topology induced as a subset of M' is a manifold.

Exercise.   Show how the Implicit Function Theorem does this job.

Definition.   This subset M defined by (5.6) is called the Pontrjagin subset of M'.

We shall now investigate the relation between this Pontrjagin subset, and the extremal field notion.

Definition.   Let (S, γ) be a pair of maps:

$$S: X \times T \to R \tag{5.7}$$

$$\gamma: X \times T \to U. \tag{5.8}$$

Let $\alpha(\gamma, S)$ be the map

$$X \times T \to M' \equiv X \times U \times Y \times T$$

defined by the following formula:

$$\alpha(\gamma, S)(x, t)$$
$$= (x, \gamma(x, t), y_i \equiv \frac{\partial S}{\partial x_i}, t) \tag{5.9}$$

<u>Definition</u>. The <u>Hamiltonian</u> (in the sense of Pontrjagin)
is the function

$$H: M \to R$$

defined as follows:

$$H = H' \text{ restricted to } M. \tag{5.10}$$

<u>Theorem 5.3</u>. If (L', γ, S) is an extremal field for the
variational problem (L, φ), then the following conditions
are satisfied:

$$\alpha(\gamma, S)(X \times T) \subset M \tag{5.11}$$

$$\frac{\partial S}{\partial t} = \alpha(\gamma, S)*(H) = 0, \tag{5.12}$$

where H is the function on M defined by (5.10). In fact,

conditions (5.11)-(5.12) are _equivalent_ to conditions
(4.9)-(4.10).

   Proof. Condition (5.11) means that

$$\alpha(\gamma,\ S)*\left(\frac{\partial H'}{\partial u_a}\right) = 0.$$

But this is precisely what is meant by (4.9). Similiarly,
from the definition (5.9) of the map $\alpha(\gamma,\ S)$, we see that
(4.10) is equivalent to (5.12).

   If the variational problem is "regular", as defined
by condition (5.4), we can see how (5.12) is related to
the traditional "Hamilton-Jacobi equation" of mathematical
physics. (See DGCV).

Theorem 5.4. If the variational problem is regular, and if
M is non-empty, then the $(x_i,\ y_i,\ t)$ restricted to M define
a coordinate system, i.e. dx_i, dy_i, dt are linearly independent.

   Proof. If the dx_i, dy_i, dt were _not_ linearly independent, there would be a non-trivial linear dependence
relation of the following form:

$$A_i dx_i + B,\ dy_i + (dt + \lambda_a \alpha\left(\frac{\partial H'}{\partial u_a}\right) = 0.$$

In view of (5.5) and (5.4), this would force:

$$\lambda_a = 0,$$

which forces $A_i = 0 = B_i = C.$

Remark: Another way of putting this is to say that:

$$(dx_i, \ dy_i, \ dt, \ d(\frac{\partial H}{\partial u_a}))$$

are linearly independent, i.e. the functions

$$(x_i, \ y_i, \ X, \ \frac{\partial H}{\partial u_a}) \qquad\qquad (5.13)$$

form (locally) a coordinate system for M'.

Exercise. Show that the coordinate system property for the functions (5.13) is equivalent to the regularity of the variational problem.

Theorem 5.5. Suppose the variational problem (L, φ) is regular, and regard the (x, y, t) as coordinates for M, so that the Hamiltonian in the sense of Pentrjagin becomes a function

$$H(x, y, t)$$

of these variables. Then, condition (5.12) is equivalent to the condition that $S(x, t)$ satisfies the Hamilton-Jacobi equation:

$$\frac{\partial S}{\partial t} = H(x, \frac{\partial S}{\partial x}, t).$$ (5.14)

Proof. The proof should be clear, if one understands the notation!

Remark: This result suggests a strategy for finding extremal fields. First, solve (5.14). Then, solve the m equations (4.9) for the m functions $\gamma(x, t) \equiv (\gamma_1(x, t),\ldots, \gamma_m(x, t))$.

Exercise. Show that this procedure does indeed provide a pair (S, γ) which defines an extremal field. Show also that the regularity of (L, φ), i.e. condition (5.4), implies that the equations can be solved for the γ_a.

Remark: To find the trajectories of the extremal field (S, γ), we shall see that it is not necessary to actually solve explicitly for γ. The trajectories will be shown to be given (as curves in X) as solutions of the Hamilton equations with Hamiltonian $H(x, y, t)$.

6. CONVEXITY CONDITIONS

Keep the notations of Section 5. We have derived the

differential equations (5.11)-(5.12) as consequences of
the extremal field conditions. We now give realistic
sufficient conditions that (5.11)-(5.12) imply the extremal
field conditions. In the classical calculus of variations,
conditions of this type are usually called Legendre conditions. I prefer to refer to them by their mathematical
property, i.e. "convexity."

Definition. The variational problem defined by the
Lagrangian L and the control system φ is said to be convex
if the Hamiltonian function

$$H(x, u, y, t) = L(x, u, t)$$
$$- y_i \varphi_i (x, u, t) \qquad (6.1)$$

is convex as a function of $(u_a) = u$ with (x, y, t) held
fixed.

Remark: If everything is differentiable (which is a usual
assumption), then convexity amounts to the following condition:

> The matrix $(\frac{\partial^2 H}{\partial u_a \partial u_b})$ is
> positive semi-definite. $\qquad (6.2)$

Theorem 6.1. Suppose (S, γ) is a pair of maps, satisfying

(5.11)-(5.12). Then, it also defines an extremal field
for the variational problem.

Proof. Set:

$$L' = \frac{\partial S}{\partial t} + \varphi_i \frac{\partial S}{\partial x_i}. \tag{6.3}$$

We have seen that the conditions (5.11)-(5.12) mean that,
for fixed (x, t), the point

$$u = \gamma(x, t)$$

is a critical point of the function on U:

$$u \to L(x, u, t) - L'(x, u, t).$$

Now,

$$L(x, \gamma(x, t)t) - L'(x, \gamma(x, t), t)$$
$$= H(x, \gamma(x, t), \frac{\partial S}{\partial x_i}, t). \tag{6.4}$$

Hence, condition (6.2) implies that:

$$u = \gamma(x, t) \text{ is an absolute minimum}$$
$$\text{of the function } u \to L(x,u,t) - L'(x,u,t) \tag{6.5}$$

Exercise. Show that if f: U → R is a convex function,
with U a real vector space, and u_0 is a critical point of
f, then

$$f(u) \geq f(u_0) \text{ for all } u \in U,$$

i.e. u_0 is absolute minimum of f.   In a formula:

$$f(u_0) = \min_{u \in U} f(u). \tag{6.6}$$

This is the result (quite easy to prove, in fact) that is
used to prove (6.5).

To complete the proof of Theorem 6.1, note that condition (5.12) implies that:

$$L(x, \gamma(x, t), t) = L'(x, \gamma(x, t), t). \tag{6.7}$$

Combining (6.7) with (6.5) proves that:

$$L(x, u, t) \geq L'(x, u, t) \tag{6.8}$$
$$\text{for \underline{all} } (x, u, t) \in X \times U \times T.$$

This proves that the Lagrangian L dominates L'; hence, as
discussed earlier, (γ, S, L') is an extremal field for L.

Now, we can discuss the relation to the "Pontrjagin
Principle."

Theorem 6.2. Suppose the variational problem (L, φ) is
convex and regular. Then,

M = set of points $(x, u, y, t) \in M'$

= $X \times U \times Y \times T$ at which the Hamiltonian function

$H'(x, u, y, t)$ takes its minimum value of u, for

fixed (x, y, t).

Further for each $(x, y, t) \in X \times Y \times T$, there is but one value u $\in$ U so that

$$(x, u, y, t) \in M,$$

i.e. M can be parameterized by $X \times Y \times R$. In these parameters,

$$H(x, y, t) = \min_{u \in U} H'(x, u, y, t). \tag{6.9}$$

Proof. The convexity plus regularity condition imply
that:

The Hessian matrix

$$(\frac{\partial^2 H'}{\partial u_a \partial u_b})$$

is positive definite, i.e. the function

$$u \to H'(x, u, y, t)$$

is strongly convex.

Exercise. Show that this implies that the function
$u \to H(x, u, y, t)$ has a unique minimum. The conclusions
follow from these facts.

Remark: Formula (6.9) may be called Pontrjagin's formula.
It is essentially what is called the Pontrjagin Maximal
Principle in the control theory and economics literature.

(Note that our "minimal" is Pontrjagin's "maximal" because his Hamiltonian is the negative of ours. I prefer to write it in this way because "convexity" is more natural to me than "concavity.")

7. THE HAMILTON EQUATIONS FOR THE EXTREMAL CURVES

Let us recapitulate. X and U are real vector spaces, called the state and control space. (They could well be chosen as manifolds, of course). T denotes a time interval. A Lagrangian

$$L: X \times U \times T \to R$$

and a control system

$$\varphi: X \times U \times T \to X$$

are given. Together, they determine the variational problem.

A function

$$S: X \times T \to R$$

defines a completely integrable Lagrangian L'. A control law is a map

$$\gamma: X \times T \to U.$$

A curve $t \to x(t)$ in X is a trajectory of the control law if:

$$\frac{dx}{dt} = \varphi(x(t), \gamma(x, t), t). \qquad (7.1)$$

A pair (S, γ) determines an <u>extremal field</u> of certain conditions are satisfied. Finally, a curve $t \to x(t)$ is an
<u>extremal curve</u> if it is a trajectory of <u>some</u> γ which is
part of some extremal field (S, γ).

This is essentially Caratheodory's definition, in his
book "Variationsrechnung". We shall now determine the
ordinary differential equations which the extremal curves
satisfy. In fact, most of this work has already been done.

Choose indices as follows:

$$1 \leq i, j \leq n = \dim X$$
$$1 \leq a, b \leq m = \dim U.$$

Suppose that:

$$x = (x_i), u = (u_a)$$

are coordinates for X and U.

The Lagrangian L, a map

$$X \times U \times T \to R,$$

becomes a function $L(x, u, t)$ of these variables. The
control system may

$$\varphi: X \times U \times T \to X$$

is defined by a set

$$\varphi_i(x, u, t)$$

of real-valued functions: $X \times U \times T \rightarrow R$.

Introduce Y, the <u>costate space</u>, and its coordinates.

$$y = (y_i).$$

Set:

$$M' = X \times U \times Y \times R$$

$$H' = L - y_i \varphi_i. \tag{7.2}$$

Now, set:

$$v_a = \frac{\partial H'}{\partial u_a}.$$

Assume that the variational problem is <u>regular</u>. Recall that this means that the functions:

$$(x, y, v, t)$$

define a new coordinate system for M'. Set:

$$\begin{array}{l} M = \text{set of points of M' on which} \\ v_a = 0. \end{array} \tag{7.3}$$

$$H = H' \text{ restricted to M} \tag{7.4}$$

(x, y, t) restricted to M then defines a coordinate system for M. H becomes a function

$$H(x, y, t)$$

of these variables.

By (5.14), the action function S, part of the extremal field (γ, S), is a solution of the "classical" Hamilton-Jacobi partial differential equation:

$$\frac{\partial S}{\partial t} = H(x, \frac{\partial S}{\partial x}, t). \tag{7.5}$$

The <u>trajectories</u> of the extremal field (γ, S) satisfy equations (7.1). Set:

$$y_i(t) = \frac{\partial S}{\partial x_i}(x(t), t). \tag{7.6}$$

$$u(t) = \gamma(x(t), t). \tag{7.7}$$

Then,

$$\frac{\partial H}{\partial y_i}(x, y, t)$$

$$= \frac{\partial H'}{\partial y_i}(x, y, v, t)/_{v=0} \quad \text{(by (7.4))}$$

$$= \text{using (7.2),}$$

$$= \varphi_i(x(t), \gamma(x(t), t), t)$$

$$=, \text{using (7.1),}$$

$$- \frac{dx_i}{dt}.$$

Hence, gathering this together we have the fundamental equations:

$$\frac{dx_i}{dt} = - \frac{\partial H}{\partial y_i}(x(t), y(t), t). \tag{7.8}$$

$$y(t) = \frac{\partial S}{\partial x_i} (x(t), t). \tag{7.9}$$

Notice that (7.8) is already one-half of the Hamilton
equations. It is a standard result of Hamilton-Jacobi
theory (see DGCV) that (7.8)-(7.9) plus the equation (7.5)
imply the field set of Hamilton equations. For completeness, we shall prove this below. First, let us sum up as
follows:

Theorem 7.1. Let (γ, S) be an extremal field of the
regular variational problem. Let $t \to x(t)$ be a curve in
X which is a trajectory of the extremal field.

$$y_i(t) = \frac{\partial S}{\partial x_i} (x(t), t). \tag{7.10}$$

Then, the curve

$$t \to (x(t), y(t))$$

is a solution of the Hamilton ordinary differential
equations:

$$\frac{dx_i}{dt} = - \frac{\partial H}{\partial y_i} (x(t), y(t), t) \tag{7.11}$$

$$\frac{dy_i}{dt} = \frac{\partial H}{\partial x_i} (x(t), y(t), t). \tag{7.12}$$

    Proof. (7.8) is identical to (7.11), so that we have

already proven (7.11). To prove (7.12), differentiate
(7.10), and use the Hamilton-Jacobi equation (7.5):

$$\frac{dy_i}{dt} = \frac{\partial^2 S}{\partial x_j \partial x_i} (x(t), t) \frac{dx_j}{dt}$$

$$+ \frac{\partial}{\partial x_i} (\frac{\partial S}{\partial t}) (x(t), t)$$

$$= - \frac{\partial^2 S}{\partial x_j \partial x_i} \frac{\partial H}{\partial y_j} + \frac{\partial}{\partial x_i} (H(x, \frac{\partial S}{\partial x}, t)) (x(t), t)$$

$$= - \frac{\partial^2 S}{\partial x_j \partial x_i} \frac{\partial H}{\partial y_j} + \frac{\partial H=}{\partial x_i} (x, \frac{\partial S}{\partial t}, t)$$

$$+ \frac{\partial}{\partial y_j} \frac{\partial^2 S}{\partial x_i \partial x_j}$$

$$= \frac{\partial H}{\partial x_i} (x(t), y(t), t),$$

which is just (7.12).

Remarks: If the variational problem is convex and regular,
this result may be used to show that the trajectories of
extremal fields satisfy the Pontrjagin Principle, summed up
in the following formulas:

$$H'(x, u, y, t) = L(x, u, t)$$
$$- y_i \varphi_i (x, u, t) \tag{7.13}$$

$$H(x, y, t) = \min_{u \in U} H'(x, u, y, t) \tag{7.14}$$

$$t \to x(t) \text{ is a curve in X which}$$
$$\text{is a projection of a curve}$$
$$t \to (x(t), y(t)) \text{ in X} \times \text{Y which} \qquad (7.15)$$
$$\text{satisfies } (7.11)\text{-}(7.12).$$

Of course, the great feature of this reformulation is that the formulas (7.13)-(7.15) <u>make sense</u> for variational problems which are more general then those with which we started. One possibility is to allow U to be more general than a manifold. For example, one might want to cover what the engineers and economists think of as <u>inequality constraints</u>. One simple possibility is to allow U to be the subset of vectors u ϵ R^m such that:

$$|u_a| \leq 1.$$

Also, formulas (7.13)-(7.15) <u>make sense</u> when the variational problem (L, φ) is <u>not</u> regular. In practice, in various applications, one then uses the equations, without worrying too much whether their solutions actually lead to rigorous minimization solutions of the variational problem.

We can also interpret these results more "geometrically", in terms of the theory of characteristic curves of 2-forms, in the way described (for ordinary variational problems) by E. Cartan in his book "Lecons sur les invariants integraux," and, for the Lagrange variational problem, in DGCV and

earlier chapters of this Volume.   I will briefly sketch
this formalism: M' is the space X × U × Y × T.

$$\theta = y_i dx_i + H'dt, \qquad\qquad (7.16)$$

with:

$$H' = L - y_i \varphi_i. \qquad\qquad (7.17)$$

Exercise. Show that the extremal curves t → x(t) in X are
the projection into X of the curves in M' which are characteristic curves of dθ. Show also that these characteristic
curves lie on the submanifold M.

Remark: A further interesting (and new) feature of this
problem is that the characteristic vector system of dθ is
singular, in the sense that its dimension is not constant
on M'. Recall (e.g. from LAQM) that the characteristic
curves play the key role in "quantization" of the system.
Then, we see that there is interesting and important new
work to be done in terms of the relation between quantum
mechanics and this sort of general variational problem. I
believe that there are important relations between this
subject and what is called "stochastic control theory." All
this I hope to get to in a later volume!

Chapter 2

HAMILTON-JACOBI THEORY AND THE
OPTIMAL CONTROL PROBLEM

1. INTRODUCTION

Hamilton-Jacobi theory is the analytical side of the calculus of variations (and classical particle mechanics). Although it is very basic to many branches of mathematics, physics and engineering, it is now rarely taught or studied for its own sake.

Of course, it can, as Cartan showed briefly (and mystifyingly, but brilliantly) in "Leçons sur les invariants intégraux", be described completely in terms of the modern (differential geometric) differential form--vector field formalism. See Abraham and Marsden and DGCV for this point of view. However, it can also be described completely within the category of vector spaces at an intermediate level of mathematical sophistication. For certain purposes, this approach is even more desirable. As a diversion from the main topics in this book, I will now present such a formalism.

I also have in mind preparing the way for a Hamilton-Jacobi theory in an infinite dimensional setting. (Physically, this involves generalizing from "particles" to "fields".) As we proceed, I will make comments that should keep this possibility in view.

Another goal of this chapter is the development of a
formalism which will be useful in optimal control theory.

2.   TAYLOR'S SERIES AND THE DIFFERENTIALS OF REAL VALUED
     FUNCTIONS

Let X be a real vector space. (Suppose it to be finite
dimensional, although the possibility that the ideas extend
to infinite dimensions should be kept in mind.) Consider a
real-valued function

$$f: X \to R \quad .$$

Suppose it to be C^{∞} in the neighborhood of a point $x_0 \in X$.

Treating X as a special sort of <u>manifold</u>,

$$df(x_0) \quad ,$$

the <u>differential of</u> f <u>at</u> x_0, is defined as usual in differ
ential geometry as a <u>tangent covector</u> to the manifold X,   i.e
an element of

$$X^d_{x_0} \quad .$$

(X_{x_0} denotes the tangent space to X at x_0. $X^d_{x_0}$ denotes
its <u>dual space</u>.)

Since X is a vector space, the tangent space X_{x_0} can
be identified with X itself. (In this identification, $x \in X$
is identified with the tangent vector to the curve

$$t \to x_0 + tx$$

at $t = 0$.) Thus, the space $X^d_{x_0}$ of one-covectors is
identified with X^d, the dual space to X. In particular,

$$df(x_0)$$

is identified with an element of X^d.

In this chapter, we shall not use the general
manifold formalism. df will denote the map

$$X \to X^d .$$

Remark. Already, we can see a connection with "symplectic
manifold" theory, in terms of material considered in applications, e.g., optimization theory. The graph of df is then
a map

$$X \to X \times X^d$$

$X \times X^d$ is a symplectic manifold. This symplectic structure
(which we have already seen play a basic role in many areas
of application) is then obviously related to the "optimization
geometry" associated with f.

Another way of defining the differential of a function
in a vector space is via Taylor's series.

$$f(x_0+x) \; = \; f(x_0) \; + \; df(x_0)(x) \; + \cdots \qquad (2.1)$$

(The terms ... denote terms of higher order.) This formula
makes it obvious why it is natural (in a vector space setting)
to consider $df(x_0)$ as an element of the dual space to X.

Formula (2.1) suggests how the <u>higher order differentials</u>
of f,

$$d^2 f(x_0), \; d^3 f(x_0), \ldots$$

are to be defined. For each integer r,

$$d^r f(x_0)$$

is an r-<u>multilinear</u>, <u>symmetric</u> map

$$\underbrace{X \times \cdots \times X}_{r \text{ copies}} \to R$$

Another way of putting it is to say that

$$d^r f(x_0) \; \varepsilon \; \underbrace{X^d \circ \cdots \circ X^d}_{r \text{ copies}}$$

($\circ$ denotes <u>symmetric tensor product</u> of vector spaces. Thus,
we identify the symmetric tensor product of r copies of X^d
with the space of r-<u>th degree polynomial functions</u> on X.)

With this concept at hand, we can write the complete
Taylor's series:

$$f(x_0+x) = f(x_0) + df(x_0)(x) + \frac{1}{2!} d^2f(x_0)(x,x)$$

$$+\cdots+ \frac{1}{r!} d^r f(x_0)(x,\ldots,x) \qquad (2.2)$$

$$+\cdots$$

We can compare this with the following one-variable Taylor's series:

$$f(x_0+tx) = \sum_{r=0}^{\infty} \frac{1}{n!} \left(\frac{d^r}{dt^r} f(x_0+tx) \right)\bigg|_{t=0} t^r \qquad (2.3)$$

This gives the following formula:

$$d^r f(x_0)(x,\ldots,x) = \frac{d^r}{dt^r} (f(x_0+tx))|_{t=0} \qquad (2.4)$$

Let us now generalize these formulas for functions $f(x,y,\ldots)$ of several vector arguments. For simplicity, restrict attention to two variables x,y. Suppose then that X,Y are vector spaces

$$f: X \times Y \to R$$

is a map.

Given $(x_0, y_0) \in X \times Y$, suppose f is C^∞ in a neighborhood of this point. Then, the classical Taylor's series gives the following formula:

$f(x_0 + tx, \; y_0 + sy)$

$$= \sum_{(n,m)} \frac{1}{n!} \frac{1}{m!} \frac{\partial^n}{\partial t^n} \frac{\partial^m}{\partial s^m} \; f(x_0 + tx, \; y_0 + sy) \Big|_{(t,s)=(0,0)} t^n s^m$$

Set:

$$\partial_x^n \partial_y^m \; f(x_0)(x,y) \;\; = \;\; \frac{\partial^n}{\partial t^n} \frac{\partial^m}{\partial s^m} \; f(x_0 + tx, \; y_0 + sy) \Big|_{(t,s)=(0,0)}$$

$$(2.5)$$

Thus, Taylor's series take the following elegant form:

$$f(x_0 + tx, \; y_0 + sy) \;\; = \;\; \sum_{n,m} \frac{1}{n!m!} \partial_x^n \partial_y^m \; f(x_0)(x,y) \qquad (2.6)$$

$\partial_x^n \partial_y^m \; f(x_0)$ is an element of

$$\underbrace{(X \circ \cdots \circ X)}_{n \text{ copies}} \otimes \underbrace{(Y \circ \cdots \circ Y)}_{m \text{ copies}}$$

It is called the partial differential.

3.   THE DIFFERENTIAL OF QUADRATIC FUNCTIONS

Keep the notation of Section 2.   X   is a vector space, say
over the real numbers as scalars.   As an exercise in the formalism (which will be useful to us in Hamilton-Jacobi theory)
I will now compute the differential of a quadratic function.

<u>Definition</u>.   A function

$$f: X \rightarrow K$$

is <u>quadratic</u> if it satisfies the following differential equation:

$$d^3 f = 0 \qquad\qquad (3.1)$$

<u>Exercise</u>.   Show that a quadratic function can be written in
the following form:

$$f(x) = c + y(x) + \beta(x,x) \qquad\qquad (3.2)$$

$$\text{for all}\ \ x \in X$$

with:

$$c \in R\ ,$$

$$y \in X^d$$

$$\beta \in X^d \circ X^d \quad (\equiv \text{space of symmetric bilinear forms on } X)$$

Remark. Of course, it is a trivial exercise to do in terms of
local coordinates. I suggest that the reader try for a coordinate-free proof. (3.2) can be rephrased to say that the
vector space of functions f which satisfy (2.1) is isomorphic
to

$$R \oplus X^d \oplus X^d \circ X^d \quad .$$

Exercise. Generalize to show that the set of n-th degree polynomial functions $f: X \to R$ defined to satisfy

$$d^{n+1}f = 0 \quad ,$$

is isomorphic to

$$R \oplus X^d \oplus \cdots \oplus \underbrace{X^d \circ \cdots \circ X^d}_{n \text{ copies}}$$

Let us now give a way to calculate the differential of a
quadratic function. Let

$$\beta: X \times X \to R$$

be a symmetric bilinear form, i.e., an element of

$$X^d \circ X^d \quad .$$

We can associate to β a linear map

$$\alpha: X \to X^d \quad ,$$

via the following formula:

$$\boxed{\begin{array}{l} \alpha(x)(x') \quad = \quad \beta(x,x') \\[2ex] \text{for} \quad x,x' \ \varepsilon \ X \end{array}} \qquad (3.3)$$

Exercise. Show that the assignment

$$\beta \rightarrow \alpha$$

defined by formula (3.3) sets up a linear map

$$X^d \circ X^d \rightarrow L(X,X^d)$$

Show that the image of this linear map is the set of

$$\alpha \ \varepsilon \ L(X,X^d)$$

which satisfy the following condition:

$$\boxed{\begin{array}{l} \alpha(x)(x') \quad = \quad \alpha(x')(x) \\[2ex] \text{for} \quad x,x' \ \varepsilon \ X \end{array}} \qquad (3.4)$$

Remark. An $\alpha \ \varepsilon \ L(X,X^d)$ which satisfies (3.4) is said to be symmetric. Denote the space of such maps by

$$LS(X,X^d) \quad .$$

Notice that in order to define "symmetry" for maps the domain and range must be dual, i.e., it makes no sense to try to define

a notion of "symmetry" for maps $\alpha: X \to Y$ between <u>arbitrary</u>
vector spaces. (Of course, this is also obvious from the poi
of view of tensor analysis. Linear maps have one "covariant"
and one "contravariant" index; it makes no sense to interchange them unless there is a "duality" relation available to
convert one type into the other.)

    We can now sum up as follows:

<u>Theorem 3.1</u>. Let $f: X \to R$ be a quadratic map. Then, there
is a $c \in R$, $y \in X^d$, $\alpha \in LS(X, X^d)$ such that:

$$f(x) = c + y(x) + \alpha(x)(x)$$

$$\text{for } x \in X$$

(3.5

    Formula (3.5) is ideal for computing the differentials
of a quadratic f. It is called the <u>polarized form</u> of f.

<u>Theorem 3.2</u>. For a quadratic function $f: X \to R$ of the form
(3.5), the differentials are computed as follows:

$$df(x_0) = y + 2\alpha(x_0)$$

$$d^2 f(x_0)(x) = 2\alpha(x)(x)$$

(3.6

Proof.

$$f(x+x_0) - f(x_0) = y(x) + \alpha(x+x_0)(x+x_0) - \alpha(x_0)(x_0)$$

$$= (y(x) + \alpha(x)(x_0) + \alpha(x_0)(x)) + \alpha(x)(x)$$

$$= \text{, using the symmetry condition (3.5),}$$

$$(y(x) + 2\alpha(x_0)(x)) + \alpha(x)(x)$$

$$= (y + 2\alpha(x_0))(x) + \alpha(x)(x)$$

The first term on the right hand side is now <u>linear</u> in x, the second <u>quadratic</u>. Taylor's series now implies formulas (3.6).

These formulas are often used (implicitly) in Applied Mathematics in optimization problems. Here is one formulation:

<u>Exercise</u>. Suppose that $x \to f(x)$ is a quadratic function of form (3.5). Show that a point $x_0 \in X$ is a critical point of f if

$$\boxed{\alpha(x_0) = -y} \tag{3.7}$$

If the homogeneous quadratic form $x \to \alpha(x)(x) \equiv d^2f(x,x)$ is positive semi-definite, show that a point x_0 satisfying (3.7) is an <u>absolute minimum</u> of f, i.e.,

$$f(x) \geq f(x_0)$$

$$\text{for all} \quad x \in X$$

(3.8)

Remark. I apologize to the mathematically sophisticated reader for the completely trivial (to him) nature of the remarks of this section. I will need this way of looking at these concepts in order to do Hamilton-Jacobi theory (and the Linear Regulator Problem of Control Theory) in a coordinate free way.

4.   THE HAMILTON EQUATIONS

Let X be a real vector space. (Suppose that it is finite dimensional, although keep in mind that generalizations to infinite dimensions are desirable.) Let Y be its dual space, i.e.,

$$Y = X^d .$$

Now, Y^d can be identified with X. We shall see that this possibility is the "algebraic" feature of Hamilton-Jacobi theory. (Of course, it generalizes to infinite dimensions in only a weak form.) To see this, map $x \in X$ into the linear map

$$y \to y(x)$$

of $Y \to R$. This maps

$$X \to (X^d)^d \quad ,$$

and <u>in finite dimensions</u> it is an isomorphism. I will usually assume, in this chapter, that this identification has been made.

A <u>Hamiltonian</u> is a map

$$H: X \times Y \to R \quad .$$

Recall from the previous section how the "partial differentials" of such an H are defined:

$$\partial_x H(x,y) \; \varepsilon \; X^d \; \equiv \; Y$$

$$\partial_y H(x,y) \; \varepsilon \; Y^d \; \equiv \; X \quad .$$

The <u>Hamiltonian equations</u> are:

$$\frac{dx}{dt} = \partial_y H(x(t),y(t))$$

$$\frac{dy}{dt} = -\partial_x H(x(t),y(t))$$

(4.1)

In Control Theory (for the Linear Regulator Problem, for example) it is important to make equations (4.1) more explicit in case H is a <u>homogeneous quadratic function</u>.

In this case, we can write H as follows:

$$H(x,y) = H_1(x) + H_2(x,y) + H_3(y) \qquad (4.2)$$

$$H_1 \ \varepsilon \ X^d \circ X^d; \quad H_2 \ \varepsilon \ X^d \otimes X^d; \quad H_3 \ \varepsilon \ Y^d \otimes Y^d \ .$$

More explicitly,

  H_1 is a homogeneous quadratic map $X \to R$

  H_2 is a bilinear map $X \times Y \to R$

  H_3 is a homogeneous quadratic map $Y \to R$

So far, we have not used the fact that Y is X^d. Let us now do so. We can "polarize" H_1, H_2, H_3 and write them in the following form:

$$
\begin{aligned}
H_1(x) &= \tfrac{1}{2} h_1(x)(x) \\[4pt]
H_2(x,y) &= h_2(x)(y) \\[4pt]
H_3(y) &= \tfrac{1}{2} h_3(y)(y)
\end{aligned}
\qquad (4.3)
$$

$$h_1 \colon X \to X^d \equiv Y$$

$$h_2 \colon X \to Y^d \equiv X$$

$$h_3 \colon Y \to Y^d \equiv X$$

are <u>linear maps</u>. h_1 and h_3 are <u>symmetric</u>.

Combining (4.1) and (4.3) enables us to write the Hamilton equations in the following form:

$$\frac{dx}{dt} = h_3(y(t)) + h_2(x(t))$$

$$\frac{dy}{dt} = -h_1(x(t)) - h_2^d(y(t))$$

(4.4)

These equations will be called the linear Hamiltonian systems.

5.   THE POISSON BRACKET

The duality between X and Y will enable us to define the Poisson bracket. Recall the following conventions:

$$Y^d = X; \quad X^d = Y .$$

For $x \in X$, $y \in Y$, let

$$\langle x,y \rangle \equiv \langle y,x \rangle$$

be the real number resulting from evaluating the linear form $y \in Y \equiv X^d$ on x. The map

$$(x,y) \rightarrow \langle x,y \rangle$$

(sometimes called a dual pairing) defines a non-degenerate, bilinear map

$$X \times Y \rightarrow R .$$

Suppose given now two functions

$$f_1, f_2: X \times Y \rightarrow R \quad.$$

Set:

$$\{f_1, f_2\}(x,y) = \left\langle \frac{\partial f_1}{\partial y}(x,y), \frac{\partial f_2}{\partial x}(x,y) \right\rangle$$

$$- \left\langle \frac{\partial f_2}{\partial y}(x,y), \frac{\partial f_1}{\partial x}(x,y) \right\rangle$$

(5.1)

As (x,y) varies over $X \times Y$ this formula defines another function on $X \times Y$, called the <u>Poisson bracket</u> of f_1 and f (Keep in mind the convention inherent in formula (5.1):

$$\frac{\partial f_1}{\partial y}(x,y) \ \varepsilon \ Y^d \ \equiv \ X$$

$$\frac{\partial f_1}{\partial x}(x,y) \ \varepsilon \ X^d \ \equiv \ Y \quad.)$$

Here are some properties of this operation; their proof is left to the reader as an exercise.

$$\{f_1, f_2\} = -\{f_2, f_1\} \tag{5.2}$$

$$\{f_1, f_2, f_3\} = \{f_1, f_2\}f_3 + f_2\{f_1, f_3\} \tag{5.3}$$

$$\{f_1, \{f_2, f_3\}\} = \{\{f_1, f_2\}, f_3\} + \{f_2, \{f_1, f_3\}\} \tag{5.4}$$

$$\{f_1, f\} = 0 \quad \text{for all} \quad f \Rightarrow f_1 = \text{constant} \qquad (5.5)$$

(5.2) (skew-symmetry) and (5.4) (the Jacobi identity) say that the Poisson bracket makes the set of all functions into a <u>Lie algebra</u>. (5.2) (the derivation identity) plays a key role in quantization.

Here is a main property:

<u>Theorem 5.2</u>. Consider Hamilton's equations:

$$\frac{dx}{dt} = \frac{\partial H}{\partial y} ; \quad \frac{dy}{dt} = - \frac{\partial H}{\partial x} \qquad (5.6)$$

A function

$$(t,x,y) \rightarrow f(x,y,t)$$

mapping $T \times X \times Y \rightarrow R$ is <u>constant</u> along all solution of (5.6) if and only if it satisfies the following equation:

$$\boxed{\frac{\partial f}{\partial t} + \{H, f\} = 0} \qquad (5.7)$$

<u>Proof</u>. Let $t \rightarrow (x(t), y(t)) \varepsilon X \times Y$ be a solution of (5.6). Then,

$$\frac{d}{dt} f(x(t), y(t), t) = f_x\left(\frac{dx}{dt}\right) + f_y\left(\frac{dy}{dt}\right) + \frac{\partial f}{\partial t}$$

$$= f_x\left(\frac{\partial H}{\partial y}\right) - f_y\left(\frac{\partial H}{\partial x}\right) + \frac{\partial f}{\partial t} \qquad (5.8)$$

This vanishes if (5.7) is satisfied. The converse is covered
by the following

Exercise. Prove that (5.7) is satisfied identically if (5.8)
vanishes for each solution of (5.6).

Exercise. If f_1, f_2 are two functions which satisfy (5.7),
show that their Poisson bracket

$$\{f_1, f_2\}$$

also satisfies (5.7).

Remark. These facts are of the greatest importance for physics.
The f's satisfying (5.7) are the conservation laws. (For
"particles", they are zero--th order differential forms, i.e.,
"functions". For "fields" in three-space dimensions, they are
three-differential forms. See IM, Vol. 5.) The Poisson bracket
defines a Lie algebra structure in the space of such conservation
laws. One goal of "quantization" (in the Schrödinger-Dirac
method) is to find suitable linear representation by skew-
Hermitian operators in a Hilbert space of certain Lie subalgebra
of this Lie algebra.

In particular, we can calculate the Poisson bracket for
homogeneous quadratic functions of (x,y). This will calculate
the Lie algebra of the linear symplectic group. Suppose:

$$H(x,y) \;=\; h_1(x)(x) \,+\, h_2(x)(y) \,+\, h_3(y)(y) \quad ,$$

here:

$$h_1: X \to Y \equiv X^d$$

$$h_2: X \to Y^d \equiv X$$

$$h_3: Y \to Y^d \equiv X$$

re linear maps. Similarly, suppose:

$$H'(x,y) = h_1'(x)(x) + h_2'(x)(y) + h_3'(y)(y) \quad .$$

ow

$$\partial_x H = 2h_1(x) + h_2^d(y)$$

$$\partial_y H = 2h_3(y) + h_2(x)$$

$$<\partial_y H, \partial_x H'> = <2h_3(y) + h_2(x), \ 2h_1'(x) + h_2'^d(y)>$$

$$= 4<h_3(y), \ h_1'(x)> + 2<h_3(y), \ h_2'^d(y)>$$

$$+2<h_2(x), \ h_1'(x)> + <h_2(x), \ h_2'^d(y)>$$

$$= <y, 4h_3 h_1'(x) + h_2 h_2'(x)>$$

$$+2<h_2' h_3(y), y> + 2<h_1' h_2(x), \ x>$$

hen

$$\{H, H'\} = <\partial_y H, \ \partial_x H'> - <\partial_y H', \ \partial_x H>$$

$$= <y, (4(h_3 h_1' - h_3' h_1) + (h_2 h_2' - h_2' h_2))(x)>$$

$$+ <y, 2(h_2' h_3 - h_2 h_3')(y)> + 2<x, (h_1' h_2 - h_1 h_2')(x)> \quad (5.9)$$

Let us interpret this as follows. Write H as:

$$H \sim h_1 \oplus h_2 \oplus h_3 \; \varepsilon \; LS(X,Y) \oplus L(X,X) \oplus LS(Y,X)$$

This identifies the Lie algebra of $Sp(n,R)$ (where n = dim (the symplectic group) as a vector space, with

$$LS(X,Y) \oplus L(X,X) \oplus LS(Y,X)$$

In terms of this isomorphism, the Lie algebra bracket is given by the following formula:

$$
\begin{aligned}
\{(h_1 &+ h_2 + h_3), (h_1' + h_2' + h_3')\} \\
&= \; 2(h_1'h_2 - h_1 h_2') \\
&\oplus \; (4(h_3 h_1' - h_3' h_1) + [h_2, h_2'] \\
&\oplus \; 2(h_2' h_3 - h_2 h_3')
\end{aligned}
$$

(5.1

6.   THE HAMILTON-JACOBI EQUATION

Continue with the following notations:

X is a real vector space

$Y \equiv X^d$

T = an interval of real numbers, parameterized by t

Suppose that

$$S: X \times T \to R$$

is a real valued function. As usual, denote its "partial differential" as

$$\partial_x S(x,t) \in X^d \equiv Y \quad .$$

Define a map

$$\phi_S: X \times T \to X \times Y$$

as follows:

$$\boxed{\phi_S(x,t) \;=\; (x, \partial_x S(x,t))} \tag{6.1}$$

Let

$$H: X \times Y \to R$$

be a Hamiltonian function.

As we have seen, we can use H to write down <u>Hamilton's</u>
<u>equations</u>:

$$\boxed{\begin{aligned} \frac{dx}{dt} &= \partial_y(H)(x(t),y(t)) \\[2em] \frac{dy}{dt} &= -\partial_x(H)(x(t),y(t)) \end{aligned}} \tag{6.2}$$

Definition. S is a solution of the Hamilton-Jacobi equation (with H as Hamiltonian) if

$$\frac{\partial s}{\partial t}(x,t) + H(x, \partial_x S(x,t)) = 0$$

(6.3)

for all $(x,t) \in X \times T$

Definition. Given a function $S: X \times T \to R$, a Hamiltonian S, the following system of ordinary differential equations are called the ray equations:

$$\frac{dx}{dt} = H_y(x(t), \partial_x S(x(t), t))$$

(6.4)

The solutions of (6.4) are called rays.

Remark. In the classical literature (e.g., Caratheodory [1]) equation (6.4) is not given a name.

Theorem 6.1. The map $\phi: X \times T \to X \times Y$ given by formula (6.1) maps rays into solutions of (6.2) if and only if S is a Hamilto Jacobi function.

   Proof. Let us suppose first that S satisfies (6.3) Let

      $t \to x(t)$

be a solution of (6.4), i.e., a ray. Set:

$$y(t) = \partial_x S(x(t), t) \quad . \tag{6.5}$$

Then, the image of the curve $t \to x(t)$ under the map ϕ is the curve $t \to (x(t), y(t))$. We must how that it satisfies equations (6.2).

The first half of (6.2) is an immediate consequence of (6.4) and (6.5). Only the second half is non-trivial. Differentiate (6.5):

$$\frac{dy}{dt} = \partial_x^2 S(x(t), t)\left(\frac{dx}{dt}\right) + \partial_x\left(\frac{\partial S}{\partial t}\right)$$

$$= \quad , \text{ using (6.3) and (6.4)},$$

$$\partial_x^2 S(x(t), t)(H_y)$$

$$= \partial_x(H(x, \partial_x S))$$

$$= -\partial_x H(x, {}_x S)$$

$$= -\partial_x H(x(t), y(t))$$

which is the second equation of (6.2).

The proof of the converse, i.e., that ϕ_S mapping rays into solutions of (6.2) implies that S is a solution of (6.3), is left as an <u>exercise</u>.

Remark. Theorem 6.1 is the analytical heart of "wave-particle"
duality inherent in Hamilton-Jacobi theory that was brought to
full fruition in quantum mechanics. It can be given a very
remarkable geometric form in terms of Cartan's theory of exterior
differential systems. On $X \times T \times T$, construct the closed
two-form:

$$\omega = d(ydx - Hdt)$$

It is readily seen that S satisfies (6.3) if and only if

$$\phi_S^*(\omega) = 0 \quad ,$$

i.e., ϕ_S is a solution submanifold of the exterior differential
system generated by ω.

It is also seen that ϕ_S is a _maximal_ solution submanifold.
In particular, the Cauchy characteristic vector fields must be
tangent to it. However, the solutions of (6.2) are just the
orbit curves of this Cauchy field, i.e., the solutions of (6.2)
which touch the submanifold $\phi_S(X \times T)$ at one point lie completely
on it. This is essentially the geometric content (at least from
Cartan's point of view) of Theorem 6.1.

7.   THE LINEARIZED HAMILTON EQUATIONS

As for any system of differential equations, we can "linear-
ize" the Hamilton equations about any particular solution. It
is convenient to have this calculation available in a coordinate
free way.

Introduce two real variables,

s   and   t .

Suppose given maps

$$(s,t) \rightarrow (x(s,t),y(s,t))$$

of $R^2 \rightarrow X \times Y$

such that <u>for each</u> s,   the curve

$$t \rightarrow (x(s,t),y(s,t))$$

satisfies the Hamilton equations (6.2).   Set:

$$\hat{x}(t) = \frac{\partial x}{\partial s}(0,t)$$

$$\hat{y}(t) = \frac{\partial y}{\partial s}(0,t)$$

(7.1)

These functions of t satisfy a linear differential equation
which we now find:

$$\frac{\partial x}{\partial t} = \partial_y H$$

$$\frac{\partial y}{\partial t} = -\partial_x H .$$

Apply $\partial/\partial s$ to both sides, and set s = 0: Here is the result:

$$\frac{dx}{dt} = (\partial_{xy}^2 H)(x(t),y(t))(x(t) + (\partial_y^2 H)(x(t),y(t))(y(t))$$

$$\frac{dy}{dt} = -(\partial_x^2 H)(x(t),y(t))(x(t)) - (\partial_{xy}^2 H)(x(t),y(t))(y(t)$$

(7.2)

These are the _linearized Hamilton equations_.

We will now use them, as Caratheodory does [1], to compute
an important invariant of the Hamilton equations, that I will
call the _Caratheodory class_. This integer plays an important
role in optimal control theory. (See my paper "Some differential
geometric aspects of Lagrange variational problems".) I will
define it from the Lie point of view, rather than as Caratheodory
originally defined it.

8.   THE EXPONENTIAL MAP DEFINED BY HAMILTON'S EQUATION

Continue with the above notation. Let $H: X \times Y \to R$
be a Hamiltonian function

> For simplicity, we suppose that all solutions of
> the Hamilton equations can be defined over the
> interval $-\infty < t < \infty$.

If this condition is not satisfied, the arguments can be modified. The assumption merely simplifies notation.

Let

$$t \rightarrow g_t$$

be the one-parameter group of diffeomorphisms of $X \times Y$ generated by the solutions of Hamilton's equations (with Hamiltonian H). In words, for a given point

$$(x_0, y_0) \ \varepsilon \ X \times Y \quad ,$$

the curve

$$t \rightarrow (x(t), y(t)) \ \equiv \ g_t(x_0, y_0)$$

is a solution of equations (6.2).

Let us pick an initial point in $X \times Y$. We can choose the notation so that it is the point $(0,0)$. For each t, let

$$\alpha_t : Y \rightarrow X$$

be the map defined by the following formula:

$$\alpha_t(y) \ = \ g_t(0,y) \quad . \tag{8.1}$$

In words, $\alpha_t(y)$ is the projection in X of the solution of (6.2) which begins, for $t = 0$, at fixed x_0, i.e., 0, with y_0 arbitrary.

Definition. α_t, as a map $Y \rightarrow X$, is called the exponential associated with the Hamiltonian H. (The name was chosen bec its definition is reminiscent of what is called the "exponenti map" in Lie group theory.)

Here is a geometric picture that goes along with this:

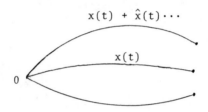

$$x(t) + \hat{x}(t)\cdots$$

$$x(t)$$

$$0$$

The curves are projections in X of solutions of (6.2). The are obtained by holding $x_0 = 0$ fixed, and varying the y_0- component.

Our goal is to compute the differential of the map α_t. Recall that, at a fixed point $y_0 \; \varepsilon \; Y$, it is a linear map $Y \rightarrow X$.

$$s \rightarrow y(s)$$

in Y. Consider the image curves

$$s \rightarrow g_t(0,y(s)) \;=\; (x(t,s),y(t,s)) \hspace{3em} (8.$$

$$s \rightarrow \alpha_t(0,y(s)) \;=\; (x(t,s)) \hspace{3em}.$$

Now, <u>for fixed</u> s,

$$t \rightarrow (x(t,s), y(t,s))$$

is a solution of Hamilton's equations. Set:

$$\hat{x}(t) = \left. \frac{\partial x}{\partial s} \right|_{s=0} \quad ; \quad \hat{y}(t) = \left. \frac{\partial y}{\partial s} \right|_{s=0} \tag{8.3}$$

We know that the curves $t \rightarrow (\hat{x}(t), \hat{y}(t))$ are solutions of the linearized Hamilton equations (7.2). The initial conditions for $t = 0$ implicit in (8.1) imply that:

$$\hat{x}(0) = 0 \quad .$$

This provides us with the following basic result:

<u>Theorem 8.1</u>. Let y_0 be a given point of Y. Let

$$(x(t), y(t)) = g_t(0, y_0) \tag{8.4}$$

This is a solution of Hamilton's equations. Use it to construct the linearized equations (7.2).

$d\alpha_t$ at $y_0: Y \rightarrow X$ is then constructed as follows:

> Given $\hat{y} \in Y$, solve equations (7.2) with initial
> values $\hat{x} = 0, \hat{y}_0$ for $t = 0$. $d\alpha_t(y_0)$ is then the
> x-component of the solution at t.

$$\tag{8.5}$$

9.   THE DIFFERENTIAL OF THE EXPONENTIAL MAP AT $y = 0$

There is one case when $y = 0$ where the solution of the
linearized Hamilton differential equations determining the
exponential map can be solved explicitly (since they have
constant coefficients). Let us examine this case in more detail.

Recall the notation. X, Y, $H(x,y)$ are as before.

$$\alpha_t(0,y) = \text{x-component of the value at time } t$$
$$\text{of the solution of Hamilton's equation}$$
$$\text{with initial condition } (0,y)$$

Now, consider the curve

$$s \rightarrow s\hat{y}_0$$

in Y. Let

$$x(t,s), \ y(t,s)$$

be the functions such that:

For fixed s,

$$t \rightarrow (x(t,s),y(t,s))$$ (9.1)

is a solution of Hamilton's equations.

$x(0,s) = 0$
$y(0,s) = s\hat{y}_0$ (9.2)
for all s

Set:

$$\hat{x}(t) = \frac{\partial x}{\partial s}(t,0)$$

$$\hat{y}(t) = \frac{\partial y}{\partial s}(t,0)$$

(9.3)

Make (9.1) explicit:

$$\frac{\partial x}{\partial t} = \partial_y H$$

$$\frac{\partial y}{\partial t} = -\partial_x H$$

(9.4)

Of course, the linearized Hamilton equations are obtained by differentiating both sides of (9.4) with respect to s, and setting s = 0. Let us do it explicitly, taking the special initial conditions (9.4) into account:

$$\frac{d\hat{x}}{dt} = \partial_x \partial_y (H)(0,0)(\hat{x}) + \partial_y^2 (H)(0,0)(\hat{y})$$

$$\frac{d\hat{y}}{dt} = -\partial_y \partial_x (H)(0,0)(\hat{y}) - \partial_x^2 (H)(0,0)(\hat{x})$$

$$\hat{x}(0) = 0$$

$$\hat{y}(0) = 0$$

(9.5)

$d\alpha_t(0): Y \to X$ is now the map which takes the initial value
in (9.5) into the value $\hat{x}(t)$ of the solution of (9.5). The
simplifying special feature is that the system (9.5) is time-
independent.

We can then write (9.5) in the following general form:

$$\frac{d\hat{x}}{dt} = A\hat{x} + B\hat{y}$$

$$\frac{d\hat{y}}{dt} = C\hat{x} + D\hat{y} \qquad\qquad (9.$$

$$\hat{x}(0) = 0; \quad \hat{y}(0) = \hat{y}_0$$

where

$$A: X \to X ; \quad B: Y \to X ;$$

$$C: X \to Y ; \quad D: Y \to X$$

are linear maps.

Theorem 9.2. Suppose that:

$$B \quad \text{is one-one} \qquad\qquad (9.$$

Then, for t sufficiently small, the map

$$d\alpha_t(0): Y \rightarrow X$$

is a vector space _isomorphism_.

Proof. Since Y and X have the same dimension, to say that B is one-one is the same as saying that it is _onto_.

Equations (9.6) can, of course, be solved in terms of power series:

$$\frac{d^2\hat{x}}{dt^2} = A\left(\frac{d\hat{x}}{dt}\right) + B\frac{d\hat{y}}{dt}$$

$$= A(A\hat{x}+B\hat{y}) + B(C\hat{x}+D\hat{y})$$

Hence,

$$\hat{x}(0) = 0$$

$$\frac{d\hat{x}}{dt}(0) = B\hat{y}_0$$

$$\frac{d^2\hat{x}}{dt^2}(0) = AB\hat{y}_0 + BD\hat{y}_0$$

$$\hat{x}(t) = tB\hat{y}_0 + \frac{t^2}{2}(AB\hat{y}_0 + BD\hat{y}_0) + \cdots$$

$$= d\alpha_t(0)(\hat{y}_0)$$

Hence,

$$\frac{d\alpha_t(0)}{t} = B + \frac{t}{2!} (AB + BD) + \cdots \qquad (9.$$

Our hypothesis (9.7) implies that:

$$\det (B) \neq 0 .$$

Hence, (9.8) implies that:

$$\det \left(\frac{d\alpha_t(0)}{t} \right) \neq 0$$

for   t   <u>sufficiently small</u>.   But,

$$\det \left(\frac{d\alpha_t(0)}{t} \right) = t^{-n} \det (d\alpha_t(0)) ,$$

where   n = dim X,   which proves Theorem 9.2.

<u>Remark</u>. This argument is very general. One can start with
X,Y arbitrary vector spaces, (9.6) an arbitrary linear const
coefficient system. (7.8) holds. Recall, from linear algebr
that the <u>rank</u> of a linear map

$$\beta: Y \to X$$

is the dimension of the image set $\beta(Y)$.   In other words, if

$$r = \text{rank } \beta ,$$

there is an r-dimensional linear subspace   Y'   of   Y   such th

$$\beta: Y' \to X$$

is one-one, but β restricted to all $(r+1)$ dimensional subspaces is not onto.

Apply this to (9.8). If

$$r = \text{rank } B \quad,$$

we see that for t sufficiently small, $d\alpha_t(0)/t$ is one-one
when restricted to an r-dimensional linear subspace of Y
(because a small perturbation of a one-one linear map is again
one-one), hence we have:

---

rank $(d\alpha_t(0)) \geq$ rank B

for t sufficiently small

We can now return to the case of a Hamiltonian system.
Consider

$$\partial_y^2 H(0,0)$$

as a symmetric linear map

$$Y \to Y^d \equiv X \quad.$$

Theorem 9.3. In this Hamiltonian case, for t sufficiently
small, the rank of $d\alpha_t(0)$ is no smaller than rank $\partial_y^2 H(0,0)$.
In particular, if $\partial_y^2 H(0,0)$ is a non-degenerate symmetric

form on Y, then

$$d\alpha_t(0): Y \to X$$

is an isomorphism. In this case, the exponential map

$$\alpha_t: Y \to X$$

is, for t sufficiently small, a <u>local diffeomorphism</u> in the neighborhood of $y = 0$.

10. THE CARATHEODORY CLASS OF AN EXTREMAL

In his work on the Lagrange variational problem, Caratheodo introduced an important integer invariant of the extremal curves I have called it (in my paper "Some differential-geometric aspects of the Lagrange variational problem") the <u>Caratheodory Class</u> of the extremal. I will now present its definition and discuss some of its geometric properties.

Let X and Y continue as real vector spaces, with Y the dual space to X, and X the dual space to Y. Let T be the real numbers. Let

$$H: X \times Y \times T \to R$$

be a Hamiltonian function.

Let

$$t \to (x(t), y(t))$$

be a solution of the Hamilton equations

$$\frac{dx}{dt} = \partial_y(H)$$

$$\frac{dy}{dt} = -\partial_x(H) \quad .$$

(10.1)

We can now form the linearized Hamilton equations:

$$\frac{d\hat{x}}{dt} = \partial_{xy}(H)(x(t),y(t),t)(\hat{x}(t) + \partial_{yy}(H)(\hat{y}(t))$$

$$\frac{d\hat{y}}{dt} = -\partial_{xx}(H)(\hat{x}(t)) - \partial_{xy}(H)(\hat{y}(t))$$

(10.2)

Caratheodory now proposed to look for the set of solutions to (10.2) which satisfy the following additional constraint:

$$\hat{x}(t) \equiv 0$$

(10.3)

This set forms a finite dimensional vector space; its dimension is called the class (in the sense of Caratheodory) of the extremal curve $t \to (x(t),y(t))$. In other words, the class is the dimension of the space of solutions of the following system of equations:

$$\partial_{yy}(H)(x(t),y(t))(\hat{y}(t)) = 0$$

$$\frac{d\hat{y}}{dt} = -\partial_{xy}(H)(x(t),y(t))(\hat{y}(t))$$

(10.4)

Now, $\partial_{yy}(H)$ is an element of $Y^d {}_\circ Y^d$, i.e., a quadrat

form on Y. We see one obvious condition which implies that

class is zero--this quadratic form is non-degenerate. Howeve

there are other possibilities.

The class number has a connection with the exponential m

Suppose the class is non-zero; there is a non-zero solution

$t \to y(t)$ of (10.4). Consider the family

$$\sigma_s: t \to (x(t), y(t)) + s(0, \hat{y}(t))$$

of curves in $X \times Y$. Each curve of the family is <u>approximate</u>

a solution of (10.1).

Let t_0 be a fixed real number. Consider the exponenti

map

$$Y \to X$$

in the neighborhood of $y(t_0)$. It then maps the curve

$$s \to y(0) + s\hat{y}(0)$$

into a curve in X. The tangent vector to this curve at s =

is then zero. In other words:

> A non-zero class implies that the exponential
> map is singular at $y(0)$, i.e., its differ-   (10
> ential is a non-invertible linear map $Y \to X$.

In case the Hamiltonian arises from a positive semi-definite
Lagrange problem, Caratheodory has proved the converse, i.e.,
the vanishing of the class implies the non-singularity of the
exponential map. To do this seems to require the whole formalism
of the calculus of variations.

11. THE HAMILTONIAN OF THE OPTIMAL CONTROL VARIATIONAL PROBLEM

Let X and U be vector spaces. Let T be the real
numbers. Suppose given maps

f: $X \times U \times T \to X$

L: $X \times U \times T \to R$.

The optimal control variational problem is to extremize the
"performance index"

$$\int L(x(t),\ u(t),\ t)\ dt \tag{11.1}$$

over curves $t \to x(t),\ u(t)$ in $X \times U$, subject to the constraint

$$\frac{dx}{dt} = f(x,u,t)\ . \tag{11.2}$$

Here is how the extremals are formed. Let Y be the dual
space of X. Define a map

$$K: X \times Y \times U \times T \rightarrow R$$

by the following formula:

$$K(x,y,u,t) = L(x,u,t) - \langle y, f(x,u,t) \rangle \ . \qquad (1$$

Now, find a map

$$\phi: X \times Y \times T \rightarrow U$$

such that:

$$\partial_\mu K(x,y,\phi(x,y,t),t) = 0 \qquad (1$$

Set:

$$H(x,y,t) = K(x,y,\phi(x,y,t),t) \qquad (1$$

The extremals of the variational problem (11.1)-(11.2) are now the projections in x-space of the solutions of the Hamil equations, with H the Hamiltonian.

We can derive several important identities by differentiating (11.4) and (11.5):

$$\partial_y(H)(x,y,t) = \partial_y K(x,y,\phi(x,y,t),t) \qquad (1$$

$$\partial_y \partial_u K(x,y,\phi(x,y,t),t) = -\partial_u^2(K)(x,y,\phi(x,y,t),t)\partial_y(\phi) \qquad (1$$

Apply ∂_y to (11.6), taking into account (11.7):

$$\partial_y^2 H(x,y,t) = \partial_y^2(K)(x,y,\phi(x,y,t),t) - \partial_u^2(K)\partial_y(\phi)\partial_y(\phi)$$

$$(1$$

However, we now see from (11.3) that:

$$\partial_y(K)(x,y,u,t) = f(x,u,t) \tag{11.9}$$

$$\partial_y^2 K = 0 \tag{11.10}$$

Substitute (11.9) and (11.10) into (11.6) and (11.8):

$$\partial_y(H)(x,y,t) = f(x,\phi(x,y,t),t) \tag{11.11}$$

$$\partial_y^2(H)(x,y,t) = -\partial_u^2(K)\partial_y(\phi)\partial_y(\phi) \tag{11.12}$$

(11.11) and (11.12) are key relations in the classical calculus of variations.

As an illustration of their usefulness, consider the solutions of the Hamilton equations, with H the Hamiltonian, i.e., the <u>extremal</u> of the variational problems (11.1)-(11.2).

$$\frac{dx}{dt} = \partial_y(H)$$
$$\tag{11.13}$$
$$\frac{dy}{dt} = -\partial_x(H) \quad .$$

In view of (11.11), we see that these relations imply that:

$$\frac{dx}{dt} = f(x,\phi(x,y,t),t) \tag{11.14}$$

This relation means that the curve $t \to x(t)$ satisfies relation (11.2), with an input curve $t \to u(t) \equiv \phi(x(t),y(t),t)$.

A map

$$W: X \times T \to R$$

is a solution of the Hamilton-Jacobi equation if:

$$\partial_t W + H(x, \partial_x W) = 0 \qquad (1$$

The family of curves in X which are the solutions of the following equations are called the <u>rays</u> associated with W:

$$\frac{dx}{dt} = \partial_y(H)(x,(t), \partial_x(W)(x(t),t), t) \qquad (1$$

In view of (11.11), we can write the ray equation (11.16) in the following alternate form:

$$\frac{dx}{dt} = f(x, \phi(x,\partial_x W,t),t) \qquad (1$$

Thus, we see that the w's which are solutions of the Hamilton-Jacobi equation (11.15), have the following propert When they are used as "feedback" in the relation

$$u = \phi(x,\partial_x w,t) ,$$

they lead to curves in X which are extremals of the variat problems (11.1)-(11.2):

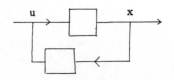

12. THE LINEAR REGULATOR PROBLEM

Continue with the notation of Section 11. Let us now specialize the variational problems (11.1)-(11.2) to the following form:

$$L(x,u) = \alpha(x)(x) + \beta(x)(u) + \frac{1}{2} \gamma(u)(u) \qquad (12.1)$$

$$f(x,u) = Ax + Bu \qquad (12.2)$$

(In these formulas, we have, for notational convenience, suppressed explicit dependence on t. The reader will find it a relatively easy task to put it back in.)

Here are the properties of α, β, γ, A, B.   X and U are real vector spaces.   Y is the dual space to X.

$$\alpha: X \rightarrow Y ; \quad \beta: X \rightarrow U^d ; \quad \gamma: U \rightarrow U^d$$

$$A: X \rightarrow X ; \quad B: U \rightarrow X .$$

are linear maps.

Following formula (11.3),

$$K = L - \langle y, Ax+Bu \rangle$$
$$= L - \langle y, Ax \rangle - \langle B^d y, u \rangle \qquad (12.3)$$

Hence,

$$\partial_u(K) = \partial_u L - B^d(y)$$

$$= \text{, using (12.1),}$$

$$\gamma(u) + \beta(x) - B^d(y) \quad .$$

Let us try to satisfy relation (11.4) with a map

$$\phi: X \times Y \to U$$

which is <u>linear</u>. Thus,

$$\phi(x,y) = \phi_1(x) + \phi_2(y) \quad , \tag{12}$$

with $\phi_1: X \to U$, $\phi_2: Y \to U$ linear maps.

With this assumed form for   , relation (11.4) takes
the following form:

$$\gamma(\phi_1(x) + \phi_2(y)) + \beta(x) - B^d(y) = 0 \quad ,$$

or:

$$\boxed{\begin{array}{rcl} \gamma\phi_1 & = & -\beta \\[2ex] \gamma\phi_2 & = & B^d \end{array}} \tag{12}$$

Assuming that γ^{-1} exists (if γ^{-1} does not exist, one says
that the variational problem is <u>singular</u>) we can then rewrite

all the formulas in the following form:

$$\phi(x,y) = \gamma^{-1}(B^d(y) - \beta(x)) \tag{12.6}$$

$$H(x,y) \equiv K(x,y,\phi(x,y))$$

$$= L(x, \gamma^{-1}(B^d(y)-\beta(x))$$

$$- \langle y,Ax\rangle - \langle y, B\gamma^{-1}(B^d y-\beta(x))\rangle$$

$$= \alpha(x)(x) + \beta(x)(\gamma^{-1}(B^d(y)-\beta(x)))$$

$$+ \frac{1}{2}(B^d(y) - \beta(x))(\gamma^{-1}B^d y - \gamma^{-1}\beta(x))$$

$$- \langle y,Ax\rangle - \langle y, B\gamma^{-1}B^d y\rangle + \langle y, B\gamma^{-1}\beta(x)\rangle$$

$$
= \boxed{
\begin{array}{l}
\langle \alpha(x),x\rangle + \frac{1}{2}\langle\beta(x),\gamma^{-1}B^d y\rangle \\[2ex]
- \frac{1}{2}\langle\beta(x),\gamma^{-1}\beta(x)\rangle \\[2ex]
+ \frac{1}{2}\langle y,B\gamma^{-1}B^d y\rangle - \frac{1}{2}\langle y,B\gamma^{-1}\beta(x)\rangle \\[2ex]
- \langle y,Ax\rangle - \langle y,B\gamma^{-1}B^d y\rangle \\[2ex]
+ \langle y,B\gamma^{-1}\beta(x)\rangle
\end{array}
} \tag{12.6}
$$

This complicated formula is one of the most important for the engineering applications. (See Anerson-Moore [1].)

13. THE MATRIX RICCATI EQUATION FOR THE LINEAR REGULATOR PROBLEM

In the Optimal Control literature a certain system of non-linear ordinary differential equations--called the matrix Ricca equations--play a key role. Usually, these equations are intr duced via a stochastic control system. The solutions of the matrix-Riccati equations are closely related to the constructio of the so-called Kalman filter. It is less well-known that the matrix Riccati equations are just a special form of our ol friend, the Hamilton-Jacobi equation. (See IM, Vol. 3.) (The real reason for this is that the stochastic control problem wh gives rise to the construction of the Kalman filter is, in a sense, "dual" to a deterministic control problem. However, th only stochastic control problems for which we understand the conceptual basis for this phenomenon are the "linear, Gaussian" processes. Luckily, these have been extremely important for applications, particularly to aerospace problems. See Anderso Moore [1] for the best available explanation of all of this. Unfortunately, the generalization of the Kalman filter to non-linear stochastic control problems has been very difficult-- there does not seem to be any analogous "duality" between non-linear stochastic and deterministic control problems.)

Let us turn to the deterministic linear regulator problems.
Let X be its state (vector) space, and let $Y \equiv X^d$ be its
dual space. Let $H(x,y)$ be the Hamiltonian for a linear regulator problem. As we have seen in Section 12, H is bilinear
on x,y. This means that it can be written in the following
form:

$$H(x,y) \ = \ \langle h_1(x),x \rangle + \langle y,h_2(x) \rangle + \langle y,h_3(y) \rangle \qquad (13.1$$

where

$$h_1: X \rightarrow Y$$
$$h_2: X \rightarrow X$$
$$h_3: Y \rightarrow X$$

are linear maps. We may then try to solve the Hamilton-Jacobi
equation by means of a quadratic function of x; i.e., one of
the following form:

$$W(x,t) \ = \ \frac{1}{2} \langle w_t(x),x \rangle \qquad\qquad (13.2$$

where, for each t,

$$w_t: X \rightarrow Y$$

is a linear map. Thus,

$$\partial_x W(x,t) \ = \ w_t(x) \ .$$

The Hamilton-Jacobi equation is:

$$\frac{\partial W}{\partial t} = H(x, \partial_x W) = 0 \qquad (13.3)$$

Relations (13.1)-(13.3) combine to give the following:

$$<\partial_t w_t(x), x> + <h_1(x), x> + <w_t(x), h_2(x)>$$

$$+ <w_t(x), h_3 w_t(x)> = 0$$

Since this is to hold for all x, we have the following:

$$\boxed{\partial_t w_t = h_1 + \frac{1}{2}(w_t h_2 + h_2^d w_t) + w_t h_3 w_t} \qquad (13.4)$$

This is the definitive Matrix Riccati equation. As we have just seen, it is just the Hamilton-Jacobi equation, with a special assumption made about the form of the solution. In IM, Vol. 3, I have shown how (13.4) is a "Lie system" with respect to the action of a Lie group, namely $Sp(n,R)$ acting on the Grassman manifold of R^{2n}; where n = dim X.

We can now readily express h_1, h_2, h_3 in terms of the data determining the linear regulator variational problem described in Section 11. Here is the form the matrix Riccati equation takes in terms of this data:

$$\boxed{\frac{\partial w_t}{\partial t} = (\alpha - \frac{1}{2}\beta^d \gamma^{-1}\beta) + \frac{1}{2}(w_t(B\gamma^{-1}\beta - A) + (\beta^d \gamma^{-1}B^d - A^d)w_t) \\ - \frac{1}{2} w_t B\gamma^{-1}B^d w_t}$$

(13.5)

Chapter 3

CHARACTERISTIC CURVES, AND THE FIRST AND SECOND VARIATION FORMULAS FOR OPTIMAL CONTROL PROBLEMS

1. INTRODUCTION

Here is the general problem. Let M be a manifold, and let T be an interval of real numbers parameterized by the real variable t.

Let θ be a one-differential form on M. For each map

$$\sigma: T \to M \quad,$$

i.e., each curve, set:

$$\underset{\sim}{\theta}(\sigma) = \int_T \sigma^*(\theta) \tag{1.1}$$

The map $\sigma \to \underset{\sim}{\theta}(\sigma)$ is called the <u>action</u> on curves, defined by θ.

Another way of defining this number is via the formula:

$$\int_T \theta(\sigma'(t)) \, dt \equiv \int_T \theta(\sigma_*(\partial/\partial t)) \, dt \tag{1.2}$$

In other words,

$$t \to \sigma'(t) \; \varepsilon \; M_{\sigma(t)}$$

is the <u>tangent vector curve to</u> σ.

Let A be a vector field on M. Let $s \to g(s)$ be the one-parameter group of diffeomorphisms of M generated by A. Transform the curve σ by the group as follows:

$$\sigma_s = g(s)(\sigma) \quad .$$

We obtain a one-parameter family

$$t \to \sigma_s$$

of curves in M.

Our goal is to calculate

$$\frac{d}{ds} \, \underset{\sim}{\theta}(\sigma_s) \Big|_{s=0} \quad ,$$

and express it in terms of an integral of the values of A on σ. This integral is called the <u>first variation formula</u>.

In case this integral vanishes, we next aim to calculate

$$\frac{d^2}{ds^2} \, \underset{\sim}{\theta}(\sigma_s) \Big|_{s=0}$$

It is called the <u>second variation formula</u>.

We shall first describe these formulas in a coordinate free way, then specialize to some cases of interest in optimal control theory. In local terms, M and θ have, in this case, the following form:

$$\theta = L(x,u,t) \, dt + y(dx - f(x,u) \, dt)$$

M   =   space of variables   (x,u,t,y)   .

Notice the following general setting for this example.

M is a fiber space over X × T.

For each point of M, the value of θ

is the pull-back of a covector in X × T.

2.   FLOWS, THE FIRST VARIATION, AND TRANSVERSALITY

Let M be a manifold, s a real parameter. A <u>flow</u>
on M is a one-parameter family

s → g(s)

of diffeomorphisms of M.

V(M) denotes the Lie algebra (under Jacobi bracket)
of <u>vector fields</u> on M. The <u>infinitesimal generator</u> of the
flow s → g(s) is the one-parameter family

s → A_s

of vector fields on M, defined by the following formula:

$$A_s(\omega) \;=\; g(s)^{-1^*} \frac{\partial}{\partial s} \, (g(s)^*(\omega)) \qquad (2.1)$$

for all differential forms ω on M.

We can now apply (2.1) to derive the first and second variation formulas. Let

$$\sigma : T \to M$$

be a curve in M, θ a one-form on M. Let $s \to g(s)$ be a flow, with infinitesimal generator $s \to A_s$. Let:

$$\sigma_s(t) = g(s)(\sigma(t)) \quad , \tag{2.2}$$

i.e.,

$$\sigma_s = g(s)\sigma$$

$$\underset{\sim}{\theta}(\sigma_s) = \int_T \sigma_s^*(\theta)$$

$$= \int_T (g(s)\sigma)^*(\theta)$$

$$= \int_T \sigma^* g(s)^*(\theta)$$

Hence,

$$\frac{d}{ds} \underset{\sim}{\theta}(\sigma_s) = \int_T \frac{\partial}{\partial s} \sigma^*(g(s)^*(\theta))$$

$$= \quad , \text{ using } (2.1),$$

$$\int_T \sigma^*(g(s)^*(A_s(\theta))$$

$$= \int_T \sigma_s^*(A_s(\theta)) \tag{2.3}$$

$$= \underbrace{A_s(\theta)}(\sigma_s) \tag{2.4}$$

This is the basic formula. It can be iterated:

$$\frac{d^2}{ds^2} \underset{\sim}{\theta}(\sigma_s) = \left(\frac{\partial}{\partial s} \underbrace{A_s(\theta)} \right)(\sigma_s) + \left(\underbrace{A_s A_s(\theta)} \right)(\sigma_s) \tag{2.5}$$

We can now use the following identity relating Lie derivative, exterior derivative and contraction:

$$A(\omega) = A \lrcorner d\omega + d(A\lrcorner \omega) \tag{2.6}$$

for all differential forms ω,

all $A \in V(M)$.

Apply (2.6) to (2.4):

$$\frac{d}{ds} \underset{\sim}{\theta}(\sigma_s) = \int_T \sigma_s^*(A_s \lrcorner d\theta) + d(\sigma_s^*(\theta(A_s)))$$

$$= \int_T \sigma_s^*(A_s \lrcorner d\theta) + \int_{\partial T} \sigma_s^*(\theta(A_s)) \tag{2.7}$$

Of course, since T is an interval of real numbers, the "boundary" ∂T, in formula (2.7), is just two points.

For example, suppose $T = \{t \varepsilon R: a \leq t \leq b\}$. Then,

$$\int_{\partial T} \sigma_s^*(\theta(A_s)) \;=\; \theta(A_s)(\sigma_s(b)) - \theta(A_s)(\sigma_s(a)) \qquad (2.8)$$

Definition. Formula (2.7) is the <u>first variation formula</u>.
A tangent vector v to a point of M is <u>transversal</u> to θ
if:

$$\theta(v) \;=\; 0 \quad .$$

A curve $s \rightarrow \phi(s)$ in M is transversal to θ if its tangent vector field

$$s \rightarrow \phi'(s) \;\varepsilon\; M_{\phi(s)}$$

is transversal for each s .

Remark. We shall see later on how this formula is related
to the usual definition of transversality in the calculus of
variations (see Caratheodory [1]) and optimal control theory
(Lee and Markus [1]).

We can now sum up as follows:

Theorem 2.1. Suppose that $\sigma: T \rightarrow M$ is a curve in M such
that

$$\sigma'(t) \;\lrcorner\; d\theta \;=\; 0 \qquad\qquad\qquad (2.9)$$

for all $t \varepsilon T$.

et $s \to \sigma_s$ be a one-parameter family of curves such that:

$$\sigma_0 = \sigma$$

For each point $a \in \partial T$, the curve
$s \to \sigma_s(a)$ is transversal to θ:

hen

$$\frac{d}{dt} \underset{\sim}{\theta}(\sigma_s)\Big|_{s=0} = 0 \quad ,$$

.e., the first variation is zero.

emark. Condition (2.9) means that σ is a characteristic
urve of $d\theta$.

. OPTIMAL CONTROL PROBLEMS

Let X, and V be manifolds. T is an interval of
eal numbers. An optimal control problem is determined by
he following data:

a) A map $L: X \times V \times T \to R$, called the Lagrangian
    or performance index.

b) A fiber preserving map

    $f: X \times V \times T \to T(X)$

    called the control law.

We shall work with coordinates:

$$(x^i), \quad 1 \geq i,j \geq n, \quad \text{for} \quad X \ .$$

$$(u^a), \quad 1 \geq a,b \geq m, \quad \text{for} \quad U \ .$$

L   determines a function

$$L(x^i, u^a, t)$$

of these variables. f   determines functions

$$f^i(x,u,t) \ .$$

Introduce another manifold Y, with variables (y_i) which are "dual" to the (x^i). (Globally, $X \times Y$ is the cotangent bundle to X.)

Set:

$$\theta \ = \ Ldt + y_i(dx^i - f^i dt) \ . \tag{3.1}$$

It is a one-form on M, the space of variables (x,y,u,t). θ is called the <u>Cartan form of the variational problem</u>.

We must compute $d\theta$. To this end, set:

$$\theta^i \ = \ dx^i - f^i dt \tag{3.2}$$

$$H \ = \ L - y_i f^i \tag{3.3}$$

(H is called the <u>Hamiltonian</u>.)

$$H_a \ = \ \frac{\partial H}{\partial u^a} \ = \ \frac{\partial L}{\partial u^a} - y_i \frac{\partial f^i}{\partial u^a} \tag{3.4}$$

Thus, we have:

$$\theta = y_i dx^i + h dt \qquad\qquad (3.5)$$

$$d\theta = dy_i \wedge dx^i + dH \wedge dt \qquad (3.6)$$

Theorem 3.1. If $\sigma: T \to M$ is a characteristic curve of then,

$$\sigma^*(H_a) = 0 \qquad\qquad (3.7)$$

Proof.

$$\sigma'(t) \lrcorner d\theta = \sigma'(t)(y_i)dx^i - \sigma'(t)(x^i)dy^i - \sigma'(t)(H)dt\, dH .$$

On the right hand side, the only terms which contain du^a are

$$\sigma^*\left(\frac{\partial H}{\partial u^a}\right) du^a \quad ,$$

hence these must vanish if σ is a characteristic curve.

Remark. We have implicitly assumed that:

$$\sigma^*(dt) \neq 0 \quad ,$$

i.e., σ can be parameterized by "t". In "singular" situations, one might encounter characteristic curves along which t was constant.

Definition. The optimal control problem is non-singular
if the one-forms

$$(dt, dx^i, dy_i, dH_a)$$

are linearly independent.

Theorem 3.2. Suppose the problem is non-singular. Let:

$$M' = \text{submanifold of } M \text{ on which } H_a = 0 \quad .$$

We can consider (x^i, y_i, t) as a coordinate system for M'
Set:

$$H'(x,y,t) = H \text{ restricted to } M' \qquad (3.8)$$

Let $t \to (x(t), y(t), u(t), t)$ be a characteristic curve
of $d\theta$. Then, it satisfies the following relation:

$$H_a(x(t), y(t), u(t), t) = 0 \qquad (3.9)$$

$$\frac{dx^i}{dt} = -\frac{\partial H'}{\partial y^i}(x(t), y(t), t)$$

$$(3.10)$$

$$\frac{dy^i}{dt} = \frac{\partial H'}{\partial x^i}(x(t), y(t), t)$$

For the proof, see Section 3, Chapter 6, Vol. 9 of IM.

Remark. Equations (3.9)-(3.10) are the "smooth" version of
the Pontrjagin Principle. (To handle "inequality" constrain

one should extend this theory to the case where U is a manifold with boundary. This can be done, and will be presented at a later point in my work.)

Theorem 3.3. Suppose $\sigma: T \to M$ is a characteristic curve of $d\theta$. Then,

$$\sigma^*(\theta^i) = 0 , \tag{3.11}$$

In particular, if

$$\sigma(t) = (x(t), y(t), u(t), t) ,$$

then

$$\frac{dx}{dt} = f(x(t), u(t), t) , \tag{3.12}$$

i.e., the projection in $X \times U$ satisfies the control law.

Proof. Use (3.1)-(3.3):

$$d\theta = \frac{\partial L}{\partial x^i} dx^i \wedge dt + \frac{\partial L}{\partial u^a} du^a \wedge dt + dy_i \wedge i + y_i d\theta^i$$

$$= \frac{\partial L}{\partial x^i} (\theta^i + f^i dt) \wedge dt + \left(H_a + y_i \frac{\partial f^i}{\partial u^a}\right) du^a \wedge dt + dy_i \wedge \theta^i$$

$$+ dy_i \wedge \theta^i + y_i \left(- \frac{\partial f^i}{\partial x^j} dx^j - \frac{\partial f^i}{\partial u^a} du^a\right) n dt$$

$$= \frac{\partial L}{\partial x^i} \theta^i \wedge dt + H_a du^a \wedge dt + dy_i \wedge \theta^i - y_i \frac{\partial f^i}{\partial x^j} \theta^j \wedge dt$$

$$\tag{3.13}$$

Now, apply

$$\sigma'(t) \lrcorner$$

to both sides of (3.13). Notice that the only term of $\sigma'(t) \lrcorner d\theta$ which contains dy_i is

$$\theta^i(\sigma'(t))dy_i \quad .$$

Hence, the condition

$$\sigma'(t) \lrcorner d\theta = 0$$

forces

$$0 = \theta^i(\sigma'(t)) \quad ;$$

which is the same as (3.11).

4.   THE LAGRANGIAN FORM OF CHARACTERISTIC EQUATIONS IN
     THE NON-DEGENERATE CASE, WITH THE DIMENSION OF CONTROLS
     EQUAL TO THE DIMENSION OF STATES

As we have seen, for closed differential forms $d\theta$ which arise from optimal control variational problems, it is quite easy to write their characteristic equations in Hamiltonian form, i.e., in terms of state and costate vectors

However, one often wants to write the equation in terms of the original state and control variables. One might call this the Lagrangian form of the characteristic equations. In this section, we derive these equations, assuming that the number of controls equals the number of states, and that a certain non-degeneracy condition is satisfied.

Keep the notation of the previous section, but suppose that

$$m = n .$$

Choose the indices a,b,... which label the control variables to be identical with the indices i,j. Thus,

$$u^i$$

are the control variables. Carry over formula (3.3):

$$H = L(x,u,t) - y_i f^i(x,u,t) .$$

Hence,

$$H_i \equiv \frac{\partial H}{\partial u^i} = \frac{\partial L}{\partial u^i} - y_j \frac{\partial f^i}{\partial u^j} \qquad (4.1)$$

Recall that, in order to calculate the characteristic equations one must set these functions H_i equal to zero. In the "state-costate" formalism, one eliminates the controls using the equation $H_i = 0$, substitutes in H to obtain H'(x,y,t), then solves the Hamiltonian equations (3.10). Our aim here is to _eliminate the costate variables_ y instead. This can be most readily done if the following condition is satisfied:

$$\det \left(\frac{\partial f^i}{\partial u^j} \right) \neq 0 \qquad (4.2)$$

<u>Definition</u>. If (4.2) is satisfied, the control law is said to be <u>non-degenerate</u>.

For the remainder of this section, we suppose that the control law is indeed non-degenerate. Let

$$h_j^i(x, u, t)$$

be the inverse matrix to $(\partial f^i/\partial u^t)$, i.e.,

$$h_k^i \frac{\partial f^k}{\partial u^j} = \delta_j^i \tag{4.3}$$

Then, the condition

$$H_i = 0 \tag{4.4}$$

are equivalent to the following conditions:

$$y_i = h_i^j \frac{\partial L}{\partial u^j} \tag{4.5}$$

Let us substitute this condition to write the Cartan form θ in the following form:

$$\theta = L dt + \left(h_i^j \frac{\partial L}{\partial u^j} \right)(dx^i - f^i dt) \tag{4.6}$$

Set:

$$h_i = h_i^j \frac{\partial L}{\partial u^j} \tag{4.7}$$

Then,

$$\theta = Ldt + h_i(dx^i - f^i dt) \quad ,$$

where h_i and f^i are functions of x, u and t.

Set:

$$\theta^i = dx^i - f^i dt$$

Then,

$$\theta = Ldt + h_i \theta^i \quad ,$$

hence,

$$d\theta = \left(\frac{\partial L}{\partial x^i} dx^i + \frac{\partial L}{\partial u^i} du^i\right) \wedge dt + dh_i \wedge \theta^i + h_i d\theta^i$$

$$(4.8)$$

$$= \frac{\partial L}{\partial x^i} \theta^i \wedge dt + \frac{\partial L}{\partial u^i} du^i \wedge dt + dh_i \wedge \theta^i + h_i d\theta^i$$

Let A be a vector field in terms of the variables (x,u,t) such that:

$$A(t) = 1 \quad , \qquad \theta^i(A) = 0$$

Then, using (4.8),

$$A \lrcorner d\theta = -\frac{\partial L}{\partial x^i} \theta^i + \frac{\partial L}{\partial u^i} A(u^i) dt - \frac{\partial L}{\partial u^i} du^i + A(h_i)\theta^i$$

$$(4.9)$$

$$+ h_i A \lrcorner d\theta^i$$

Now,

$$A \lrcorner d\theta^i = -A \lrcorner (df^i \wedge dt)$$

$$= -A(f^i)dt + df^i$$

Hence, the condition that A is a <u>characteristic vector</u> <u>field of</u> $d\theta$ can be written as follows:

$$0 = \left(-\frac{\partial L}{\partial x^i} + A(h_i) + \frac{\partial f^i}{\partial x^j}\right)\theta^j + \left(\frac{\partial L}{\partial u^i} A(u^i) + \frac{\partial f^i}{\partial x^j} f^j + \frac{\partial f^j}{\partial t} - A(f^i)\right)$$

$$- \left(\frac{\partial L}{\partial u^i} - h_j \frac{\partial f^j}{\partial u^i}\right) du^i \quad ,$$

or:

$$A(h_i) = \frac{\partial L}{\partial x^i} + \frac{\partial f^i}{\partial x^j} \tag{4.10}$$

$$\frac{\partial L}{\partial u^i} A(u^i) + \frac{\partial f^i}{\partial x^j} f^j + \frac{\partial f^i}{\partial t} - A(f^i) = 0 \tag{4.11}$$

$$\frac{\partial L}{\partial u^i} = h_j \frac{\partial f^i}{\partial u^i} \tag{4.12}$$

<u>Theorem 4.1.</u> Equations (4.12) are already a consequence of Equations (4.3) and (4.7).

Proof.

$$h_j \frac{\partial f^j}{\partial u^i} \;=\; h_j^k \frac{\partial L}{\partial u^k} \frac{\partial f^j}{\partial u^i}$$

$$=\; , \text{ using } (4.3) \; ,$$

$$\frac{\partial L}{\partial u^k} \delta_i^k$$

$$=\; \frac{\partial L}{\partial u^i} \; ,$$

which is part of (4.12).

Hence, we are left with (4.10) and (4.11) as the equations of the characteristic vector fields of $d\theta$. Now, the orbit curves of the characteristic vector fields are the characteristic curves of $d\theta$. Hence, we have the following.

Theorem 4.2. A curve $t \rightarrow (u(t), x(t))$ in (control × state)-space is a characteristic curve of $d\theta$ (hence an extremal curve of the variational problem) if and only if the following conditions are satisfied:

$$\frac{d}{dt} \left(h_i(x(t),u(t),t) \right) \;=\; \frac{\partial L}{\partial x^i} + \frac{\partial f^i}{\partial x^j} \quad (x(t),u(t),t) \qquad (4.13)$$

$$\frac{\partial L}{\partial u^i} \frac{d}{dt} \left(u^i(t) \right) \;=\; -\frac{\partial f^i}{\partial x^j} f^j + \frac{\partial f^i}{\partial t} - A(f^i)(x(t),u(t),t) \qquad (4.14)$$

Example. The "ordinary" variational problem.

In this case,

$$f^i = u^i \tag{4.15}$$

Hence,

$$\frac{\partial f^i}{\partial u^j} = \delta^i_j \ ,$$

$$h^j_i = \delta^j_i \ ,$$

$$h_i = \frac{\partial L}{\partial u^i} \ ,$$

$$\frac{\partial f^i}{\partial x^j} = 0 \ .$$

Hence, (4.13) takes the form:

$$\frac{d}{dt}\left(\frac{\partial L}{\partial u^i}\right) = \frac{\partial L}{\partial x^i} \ , \tag{4.16}$$

which is the classical Euler-Lagrange equation. Equation (4.14)
is the conservation of energy identity, which follows from
(4.15).

Thus, we see that, in this case, the characteristic
equations are a generalization of the classical Euler-
Lagrange equations of the "ordinary" calculus of variations
problem.

The next case one might consider would be where the number of controls is less than the number of states. If one assumes that the problem is <u>non-degenerate</u>, in the sense:

$$\text{rank of the matrix} \quad \left(\frac{\partial f^i}{\partial u^a}\right) = m$$

In this case, one can use the equations

$$H_a = 0$$

to solve for m of the costate variables in terms of the controls u and $(n-m)$ of the remaining costate variables. An example of such a calculation is given in Section 6, Chapter 5, Vol. 9 of IM; namely, the calculation of the Euler-Lagrange equations of higher-derivative Lagrangians.

5.   THE SECOND VARIATION

Let us return to the general situation of Section 2. Suppose $\sigma: T \to M$ is a characteristic curve of $d\theta$.

Suppose $s \to g(s)$ is a flow on M, with:

$$\sigma_s = g_s \sigma , \tag{5.1}$$

$s \to A_s$ the infinitesimal generator of
            the flow.

Let us suppose that the deformation of σ defined by (5.1) has <u>fixed end points</u>. This means that:

$$A_s(\partial T) = 0 \qquad (5.2)$$

for all s.

Using (2.5), we can then write:

$$\frac{d^2}{ds^2} \int_T \sigma_s^*(\theta) = \int_T \sigma_s^* \left(\left(\frac{\partial}{\partial s} A_s \right)(\theta) \right) + \int_T \sigma_s^*(A_s A_s(\theta)) \qquad (5.3)$$

Set $s = 0$, and use the fact that $\sigma_0 \equiv \sigma$ is a characteristic curve of $d\theta$. Then, the first term on the right hand
side of (5.3) vanishes, since:

$$\int_T \sigma^* \left. \frac{\partial}{\partial s} (A_s) \right|_{s=0} (\theta) = \qquad , \text{ using } (5.2)$$

$$\int_T \left(d\theta \left(\sigma'(t), \frac{\partial}{\partial s} (A_s)(\sigma(t)) \right) \right) dt$$

which equals zero, since $\sigma'(t) \lrcorner d\theta = 0$.

Set:

$$A = A_0 \quad .$$

We then have the following formula:

$$\frac{d^2}{ds^2} \left. \int_T \sigma_s^*(\theta) \right|_{s=0} = \int_T \sigma^*(A(A(\theta))) \qquad (5.4)$$

Formula (5.4) is the <u>second variation formula for variations</u> <u>with fixed end-points</u>. (There is a slightly more complicated formula for variations whose end-points move in a "transversal" way.)

We want to study the integrand of (5.4) in more detail.

$$A(A(\theta)) \;=\; A(d(A \lrcorner\, \theta) + A \lrcorner\, d\theta)$$

$$=\; d(A \lrcorner\, A(\theta)) + A \lrcorner\, A(d\theta)$$

$$=\; d(A \lrcorner\, A(\theta)) + A \lrcorner\, d(A \lrcorner\, d\theta)$$

Hence,

$$\frac{d^2}{ds^2} \int_T \sigma_s^*(\theta)\Big|_{s=0} \;=\; \int_T \sigma^*(A \lrcorner\, d(A \lrcorner\, d\theta)) \qquad (5.5)$$

<u>Remark</u>. In deriving (5.5), we have again used Stokes' formula and the fact that A vanishes on ∂T.

This is the most useful version of the second variation formula. The following definition isolates the main geometric object involved in this formula.

<u>Definition</u>. The map

$$S: V(M) \times V(M) \to R$$

defined by the following formula:

$$S(A,B) \;=\; \frac{1}{2} \int_T \sigma^*(A \lrcorner \, d(B \lrcorner \, d\theta) + B \lrcorner \, d(A \lrcorner \, d\theta)) \qquad (5.6)$$

is called the second variation form.

Remark. One can, in fact, show that, as a consequence of the hypothesis that σ is a characteristic of $d\theta$, this for only depends on the values taken by the vector fields A,B on the curve σ, and not on their derivatives. Hence, it is in some sense a "tensorial" formula.

6.   THE SECOND VARIATION FORMULA FOR THE ORDINARY
      VARIATIONAL PROBLEM

In particular, it is difficult to extract useful information from the second variation formula, in its general form (5.4). As an illustration of the sort of calculations that are necessary, we consider the very classical case, where the "controlled" vector is the "velocity" of the state vector x, i.e.,

$$u \;=\; \dot{x}$$

$$\frac{dx}{dt} \;=\; u \quad .$$

M can then be chosen as the space of variables

$$(x^i, \dot{x}^i, t) ,$$

with:

$$\theta = Ldt + \frac{\partial L}{\partial \dot{x}^i} \theta^i \qquad (6.1)$$

$$\theta^i = dx^i - \dot{x}^i dt \qquad (6.2)$$

Set:

$$\eta_i = d\left(\frac{\partial L}{\partial \dot{x}^i}\right) - \frac{\partial L}{\partial x^i} dt \qquad (6.3)$$

Then,

$$d\theta = \eta_i \wedge \theta^i \qquad (6.4)$$

Let $\underline{P}$ be the Pfaffian system spanned by the η_i and θ^i. Then, a curve

$$t \rightarrow x(t)$$

is an extremal of the variational problem if and only if the curve

$$t \rightarrow \left(x(t), \frac{dx}{dt}, t\right)$$

is an integral submanifold of $\underline{P}$, i.e., the forms θ^i, η_i are zero when restricted to the curve.

Let A be a vector field on $(x, \dot{x}, t)$-space of the following form:

$$A = A^i(x) \frac{\partial}{\partial x^i} + \frac{\partial A^i}{\partial x^j} \dot{x}^j \frac{\partial}{\partial \dot{x}^i} \qquad (6.5)$$

(Thus, A is the prolongation of the vector field

$$A^i(x) \frac{\partial}{\partial x^i} \qquad .)$$

We have the following relation:

$$A(\theta^i) \ \varepsilon \ \underline{P} \qquad (6.6)$$

Now,

$$A \lrcorner d\theta = \eta_i(A)\theta^i - \theta^i(A)\eta_i$$

$$d(A \lrcorner d\theta) = d(\eta_i(A)) \wedge \theta^i + \eta_i(A)d\theta^i - d\theta^i(A) \wedge \eta_i - \theta^i(A)d$$

$$A \lrcorner d(A \lrcorner d\theta) = -\theta^i(A)d(\eta_i(A)) + \eta_i(A)A \lrcorner d\theta^i$$

$$+ \eta_i(A)d(\theta^i(A)) - \theta^i(A)A \lrcorner$$

$$+ \cdots$$

(The terms $\cdots$ belong to $\underline{P}$.) Now,

$$d(\theta^i(A)) + A \lrcorner d\theta^i = A(\theta^i) \quad ,$$

which is in $\underline{P}$, using (6.6). Similarly,

$$d(\eta_i(A)) + A \lrcorner d\eta_i = A(\eta_i) \quad .$$

Thus, we have:

$$A \lrcorner \ d(A \lrcorner \ d\theta) \ = \ -\theta^i(A)A(\eta_i) + \cdots \quad ,$$

hence:

$$\text{Second variation formula} \ = \ - \int_T \sigma^*(\theta^1(A)A(\eta_i)) \qquad (6.7)$$

We can make this formula more explicit, using (6.5) and (6.3)

$$\theta^i(A) \ = \ A^i(x) \qquad (6.8)$$

$$A(\eta_i) \ = \ d\left(\frac{\partial^2 L}{\partial x^j \partial \dot{x}^i} A^j + \frac{\partial^2 L}{\partial \dot{x}^j \partial \dot{x}^i} \frac{\partial A^j}{\partial x^k} \dot{x}^k \right)$$

$$\qquad - \left(\frac{\partial^2 L}{\partial x^j \partial x^i} A^j + \frac{\partial^2 L}{\partial \dot{x}^j \partial x^i} \frac{\partial A^j}{\partial x^k} \dot{x}^k \right)$$

Suppose that:

$$\sigma(t) \ = \ \left(x(t), \frac{dx}{dt} , t \right) \quad .$$

Then,

Second variation formula

$$= \ - \int_T (A^i(x(t)) \ \frac{d}{dt} \left(\frac{\partial^2 L}{\partial x^j \partial \dot{x}^i} A^j + \frac{\partial^2 L}{\partial \dot{x}^j \partial \dot{x}^i} \frac{\partial A^j}{\partial x^k} \frac{dx^k}{dt} \right)$$

$$\qquad - A^i \left(\frac{\partial^2 L}{\partial x^j \partial x^i} A^j + \frac{\partial^2 L}{\partial \dot{x}^j \partial x^i} \frac{\partial A^j}{\partial x^k} \frac{dx_k}{dt} \right)$$

= , after integrating by parts and using the fact that $A^i = 0$ at the end-point of the interval T,

$$\int_T \left[\frac{d}{dt} A^i(x(t)) \left(\frac{\partial^2 L}{\partial x^j \partial \dot{x}^i} A^j + \frac{\partial^2 L}{\partial \dot{x}^j \partial \dot{x}^i} \frac{d}{dt} A^j(x(t)) \right) \right.$$

$$\left. - A^i \left(\frac{\partial^2 L}{\partial x^j \partial x^i} A^j + \frac{\partial^2 L}{\partial \dot{x}^j \partial x^i} \frac{d}{dt} A^j(x(t)) \right) \right] \ dt$$

$$= \int_T \left[\frac{\partial^2 L}{\partial \dot{x}^j \partial \dot{x}^i} \left(\frac{d}{dt} A^i(x(t)) \right) \left(\frac{d}{dt} A^j(x(t)) \right) \right.$$

$$\left. - \frac{\partial^2 L}{\partial x^j \partial x^i} A^i(x(t)) A^j(x(t)) \right] \ dt \qquad (6$$

Formula (6.9) is the traditional form of the <u>second variation formula</u>.

<u>Exercise</u>. From formula (6.9), derive the following result:

If $t \to x(t)$ is a curve that realizes a <u>minimum</u> of $\int_T L(x(t), \frac{dx}{dt}, t) \ dt$, among all curves with the same end-points, then the quadratic form

$$\left(\frac{\partial^2 L}{\partial \dot{x}^j \partial \dot{x}^i} \left(x(t), \frac{dx}{dt}, t \right) \right)$$

is a <u>positive definite</u>.

(This condition is called the Legendre condition.)

Chapter 4

THE WEIRSTRASS CONDITION AND THE PONTRJAGIN PRINCIPLE

1. INTRODUCTION

It is well-known to experts in control theory that the Pontrjagin principle reduces to the classical Weirstrass condition in case the optimal control problem reduces to the standard "ordinary" problem of the classical calculus of variations. I will now describe this topic from the geometric point of view with which I am describing control theory and mechanics.

2. THE ORDINARY VARIATIONAL PROBLEM

Suppose (for simplicity) that X is a real vector space of dimension n. x denotes a vector of X. (x^i), $1 \leq i,j \leq n$ denotes linear coordinates on X.

Let $\dot{X}$ denote another vector space, which is isomorphic to X. It may be thought of as the <u>velocity space</u>.

The classical "ordinary" variational problem is to extremize

$$\int L\left(x(t), \frac{dx}{dt}\right) dt \qquad (2.1)$$

109

We can readily convert this into an optimal control varia
tional problem. Let X be the <u>state space</u>, and let the input
space U be identified with the velocities $\dot{X}$. Thus, the
problem takes the following form:

Extremize

$$\int L(x(t), u(t)) \, dt \qquad\qquad (2.2)$$

subject to the constraint

$$\frac{dx}{dt} = u \quad . \qquad\qquad (2.3)$$

Let Y be the dual space to X. Denote the linear coordinates of Y dual to the x^i by (y_i).

The Cartan form of the variational problems $(2.2)-(2.3)$
is given as follows:

$$\theta = Ldt + y_i(dx^i - u^i dt) \qquad\qquad (2.4)$$

$$= y_i dx^i + hdt \qquad\qquad (2.5)$$

with:

$$H = L - y_i u^i \quad . \qquad\qquad (2.6)$$

Let M be the space of the variables (x,y,u). (Global
M is the fiber space above X, whose fiber above the point
$x \in X$ is $X_x + X_x^d$. Thus, M is the <u>direct sum of the tangen
and cotangent bundles to</u> X.)

Definition. A curve $t \rightarrow (x(t), y(t), u(t))$ in M satisfies the Pontrjagin principle if the following conditions are satisfied:

a) The curve is a characteristic curve of $d\theta$, and satisfies (2.3).   (2.7)

b) For each value of t, each $u \in U$,
$$H(x(t), y(t), u(t)) \le H(x(t), y(t), u) \qquad (2.8)$$

Let us now show how condition (2.8) reduces to the Weirstrass condition. To do this, we suppose that condition (2.7) implies the following conditions:

$$y_i(t) = \frac{\partial L}{\partial u^i} \left(x(t), \frac{dx}{dt} \right) \qquad (2.9)$$

((2.9) will be satisfied if the variational problem is non-degenerate, in the sense that $\det (\partial^2 L / \partial u^i \partial u^j) \ne 0$.)

Let us now substitute (2.6) and (2.9) into the inequality (2.8):

$$L \left(x(t), \frac{dx}{dt} \right) - \frac{\partial L}{\partial u^i} \left(x(t), \frac{dx}{dt} \right) \frac{dx^i}{dt}$$

$$\le L(x(t), u) - \frac{\partial L}{\partial u^i} \left(x(t), \frac{dx}{dt} \right) u^i \qquad (2.10)$$

for all $u \in U$.

Condition (2.10) is the classical <u>Weirstrass condition</u>.

3.   THE WEIRSTRASS E-FUNCTION

We now interpret condition (2.10) in terms called, in the classical literature (e.g., see Bolza [1]) the <u>E-function</u>. ("E" stands for "excess", I believe!)

Given the Lagrangian L, a function on $X \in U \in T$ (with U identified with $\dot{X}$, in this case). The E-function is defined by the following formula:

$$E(x,u,v,t) = L(x,u,t) - L(x,v,t) - \frac{\partial L}{\partial u^i} (u^i - v^i) \qquad (3.1)$$

for $u,v \in U$, $x \in X$, $t \in T$.

E is thus a function on $X \times U \times U \times T$. (From the global point of view, we identify $X \times U$ with $T(X)$, hence $X \times U \times U \times$ is identified with $(T(X) \oplus T(X)) \times T)$. We now immediately combine (2.10) and (3.1) to give the following result:

<u>Theorem 3.1</u>. If the curve $t \to x(t)$ in X satisfies the Pontrjagin principle, then:

$$E\left(x(t), \frac{dx}{dt}, u, t\right) \leq 0 \qquad (3.2)$$

for all $u \in U$, all $t \in T$.

The result makes precise the sense in which the Pontrjagin principle is a _generalization_ of the classical Weirstrass condition. We now formulate a general version of the Weirstrass condition.

4. THE WEIRSTRASS CONDITION FOR AN OPTIMAL CONTROL PROBLEM

Adopt the standard set-up for control theory. X is a real vector space, the _state space_. U is another real vector space, the _control space_. T is an interval of real numbers. $Y \equiv X^d$ is the dual space to X, called the _costate space_. The control problem is to minimize

$$\int L(x,u,t)\, dt$$

subject to the constraints

$$\frac{dx}{dt} = f(x,u,t) \quad .$$

Set:

$$H(x,y,u,t) = L(x,u,t) - yf(x,u,t) \tag{4.1}$$

H is the _Hamiltonian_.

Let $t \to (x(t),\, y(t),\, u(t))$ be a curve in $X \times Y \times U$. The _Pontrjagin principle_ requires that the following conditions be satisfied:

For all $t \in T$,   all $u \in U$,

$$H(x(t),y(t),u(t),t) \leq H(x(t),y(t),u,t) \qquad (4.2)$$

Let us also assume that $u(t)$ lies in the <u>interior</u> of U.
Then, (4.2) implies the following equations:

$$\frac{\partial L}{\partial u}(x(t), u(t), t) = y(t) \frac{\partial f}{\partial u}(x(t), u(t), t) \qquad (4.3)$$

Suppose U has dimension m. Choose indices as follows:

$$1 \leq a,b \leq m$$

Let

$$(u^a)$$

be coordinates for U. (4.3) then means that:

$$\frac{\partial L}{\partial u^a} = y_i \frac{\partial f^i}{\partial u^a} \qquad (4.4)$$

We now want to solve these equations, to eliminate y.
We suppose that the problem is "non-degenerate", in the sense
that the rank of the matrix

$$\left(\frac{\partial f^i}{\partial u^a}\right)$$

is <u>maximal</u>.

Case 1. $m = \dim U = \dim X = n$.

Redefine indices (a) so that $a = j$. Non-degeneracy
then means that:

$$\det\left(\frac{\partial f^i}{\partial u^j}\right) \neq 0 \tag{4.5}$$

Let h^i_j be the inverse matrix to $(\partial f^i/\partial u^j)$, i.e.,:

$$h^i_j \; \frac{\partial f^j}{\partial u^k} \; = \; \delta^i_k \tag{4.6}$$

We can then solve (4.4) for y:

$$y_i \; = \; h^j_i \; \frac{\partial L}{\partial u^j} \tag{4.7}$$

Insert (4.7) into (4.1), with the following result:

$H(x(t), y(t), u, t) \; = \;$, after substitution of (4.7) for $y(t)$

$$L(x(t), u, t) \; - \; \left(h^j_i \; \frac{\partial L}{\partial u^j}\right) (x(t), u(t), t) f^i(x(t), u, t) \tag{4.8}$$

Warning. In understanding (4.8), keep in mind that $u(t)$
and u are different vectors of L. $u(t)$ is the minimum of
the function

$$u \rightarrow H(x, y, u, t) \quad,$$

with (x, y, t) held fixed.

Using (4.8) we can rewrite the basic inequality (4.2) as follows:

$$L(x(t), u(t), t) - \left(h_i^j \frac{\partial L}{\partial u^j}\right) (x(t), u(t), t) f^i(x(t), u(t), t)$$

$$\leq L(x(t), u, t) - \left(h_i^j \frac{\partial L}{\partial u^j}\right)(x(t), u(t), t) f^i(x(t), u, t) \quad .$$

This means the following

Definition. The Weirstrass E-function for the variational problem is defined by the following formula:

$$E(x,u,v,y) = L(x,u,t) - L(x,v,t) - \left(h_i^j \frac{\partial L}{\partial u^j}\right)(x,u,t)$$

$$\times (f^i(x,u,t) - f^i(x,v,t)) \tag{4.9}$$

for $x \in X$; $u,v \in U$; $t \in T$.

A curve $t \rightarrow (x(t),u(t))$ satisfies the Weirstrass condition in case:

$$E(x,u(t),v,t) \leq 0 \tag{4.10}$$

for all $v \in U$.

Of course, by the way we have constructed the E-function, the condition (4.10) is equivalent to the Pontrjagin condition (4.2).

Case 2. $\boxed{\text{m = dim U < dim X = n}}$

"Non-degeneracy", in this case, requires that the rank of the matrix

$$\left(\frac{\partial f^i}{\partial u^a}\right)$$

be m. By rearranging the ordering of rows and columns of this matrix, we can arrange that:

$$\det\left(\frac{\partial f^b}{\partial u^a}\right) \neq 0 \quad .$$

Hence, we can solve relations (4.4), giving y_i in terms of (x,u,t). Everything now proceeds as in Case 1.

Case 3. $\boxed{\text{m = dim U > dim X = n}}$

It is not clear to me that there _is_ a reasonable analogue of the Weirstrass condition in this case!

Chapter 5

AN OPTIMAL-CONTROL DISTRIBUTED-PARAMETER
AND FIELD-THEORETIC VARIATIONAL PROBLEM

1.   INTRODUCTION

In previous work I have carried out two broad programs:
First, set up the usual "lumped parameter" optimal control
variational problem in terms of manifold theory and differential forms (see Vol. IX of IM); second, describe the
usual physicist's field theory variational problem in terms
of manifold theory and differential forms (see LAQM, VB, GPS).
In this chapter I will attempt to unify these two streams
of work.

Notice that the engineering problem being treated
here is the "optimal control of distributed parameter systems".
This is an aspect of engineering science which has been much
less developed on the geometric side than the previous works.
Thus, I hope that my efforts here may lay the foundation for
the development of geometric methods of analysis of these
systems. On the physics side, the methods should be useful
in both classical and quantum field theories.

This chapter presents a brief introduction to a general
theory that I hope to return to later in much more detail.

2.   A REVIEW OF THE GEOMETRIC FIELD-THEORETIC CALCULUS
      OF VARIATIONS FORMALISM

Let X and Z be manifolds. X is the <u>independent</u>
<u>variable manifold</u>, Z is the <u>dependent variable manifold</u>.
Let coordinates on X be denoted as

$$(x^i) \quad , \qquad 1 \leq i,j \leq n = \dim X$$

Let coordinates on Z be denoted as

$$(z^a) \quad , \qquad 1 \leq a,b \leq m = \dim Z \quad .$$

Let $M(X,Z)$ denote the space of maps

$$\underline{z}: X \to Z \quad .$$

Let $M^1(X,Z)$ denote the space of first order mapping
elements. The Lie coordinates for $M^1(X,Z)$ will be denoted
as

$$(x^i, z^a, z^a_i) \quad .$$

Thus, if $\underline{z} \in M(X,Z)$, and if:

$$(\partial^1 \underline{z})^*(z^a) = f^a(x) \quad ,$$

then:

$$(\partial^1 \underline{z})^*(z^a_i) = \frac{\partial}{\partial x^i} (f^a) \quad . \tag{2.1}$$

Let

$$\omega^a = dz^a - z^a_i \, dx^i \tag{2.2}$$

e the contact forms. Then,

$$(\partial^1 \underline{z})^*(\omega^a) = 0 \qquad (2.3)$$

Let dx denote a fixed volume element differential form on X.

A <u>Lagrangian</u> is a mapping

$$L: M^1(X,Z) \to R \ .$$

n Lie coordinates (x^i, z^a, z_i^a), it is a function

$$L(x, z, z_i^a)$$

f these variables. Suppose one is given.

The <u>Cartan form</u> is an n-degree differential form $\theta(L)$ on $M^1(X,Z)$ which is intrinsically attached to L. It is defined by the following formula:

$$\theta(L) = L dx + L_a^i \omega^a \wedge \omega_i \ , \qquad (2.4)$$

where:

ω^a are the contact-forms, defined by (2.2)

ω_i are defined as follows:

$$\omega_i = \frac{\partial}{\partial x^i} \,\lrcorner\, dx \qquad (2.5)$$

$_a^i$ are functions on $M^1(X,Z)$ defined as follows:

$$L_a^i = \frac{\partial}{\partial z_i^a}(L) \qquad (2.6)$$

In LAQM, VB and GPS I have explained how the Cartan form
is used to describe the basic features of the variational
problem in a coordinate-free way.

Remark. Notice that this is the first time I have used the
classical tensor analysis upper-lower index convention for
these formulas. It is much better for calculational
purposes than the notation I have used previously.

3.   VARIATIONAL PROBLEMS WITH CONSTRAINTS

       Keep the notation of Section 2. Introduce another set
of indices as follows:

$$1 \leq u \leq p$$

Let f^u be a set of functions on $M^1(X, Z)$.

       Consider the problem of finding a map

$$\underline{z}: X \rightarrow Z$$

which extremizes the action

$$\int_X L(x, \underline{z}(x), \partial \underline{z}(x)) \, dx \quad , \tag{3.1}$$

subject to the constraint:

$$f^u(\partial \underline{z}(x)) = 0 \tag{3.2}$$

       To handle this in the usual Lagrange multiplier way,
introduce new variables

$$(\lambda_u) \quad .$$

They are called <u>Lagrange multipliers</u>. Set:

$$L' = L + \lambda_u f^u \quad . \tag{3.3}$$

Consider L' as a Lagrangian, for independent variables (x') and dependent variables (z^a, λ_u). Let θ' be the Cartan forms of this Lagrangian.

Since L' given by (3.3) does not contain the <u>derivatives</u> of the variables λ_u, we have:

$$\theta' = \theta(L) + \lambda_u \theta(f^u) \tag{3.4}$$

This is the basic formula for determining the extremals of the variational problem. Let Λ be the space of the variables (λ_u). One looks for a mapping

$$\phi: X \to Z \times \Lambda$$

such that:

$$(\partial^1 \phi)^*(A \lrcorner d\theta') = 0 \tag{3.5}$$

for all vector fields A on $M^1(X, Z \times \Lambda)$,

and such that:

$$(\partial^1 \phi)^*(f_u) = 0 \tag{3.6}$$

<u>Remark</u>. The f_u are functions on $M^1(X, Z)$. $\theta(f^u)$ denotes their Cartan form.

The "optimal control" variational problem is now a special case of the constrained variational problem described above.

4.   AN OPTIMAL CONTROL VARIATIONAL PROBLEM

Keep the previous notation. In addition to the spaces already considered, introduce another manifold U, called the control space. Suppose given functions

$$h_i^a(x,z,u) \quad , \tag{4.1}$$

$$L(x,z,u) \quad . \tag{4.2}$$

We now consider the problem of extremizing

$$\int_X L(x,\underline{z}(x), \underline{u}(x)) \, dx \quad , \tag{4.3}$$

subject to the constraint:

$$\frac{\partial z^a}{\partial x^i} = h_i^a(x,z(x), \underline{u}(x)) \tag{4.4}$$

This is the optimal control variational problem.

To reduce this problem to the constrained problem considered in Section 3, replace

$$Z \quad \text{by} \quad Z \times U \quad .$$

Set:

$$f_i^a = z_i^a - h_i^a(x,z,u) \quad . \tag{4.5}$$

The Lagrange multiplier variables will then be denoted by:

$$\lambda_a^i \tag{4.6}$$

Set:

$$L' = L + \lambda_a^i f_i^a \quad . \tag{4.7}$$

Let

$$\theta_i^a$$

be the Cartan form of f_i^a. We have;

$$\theta_i^a = f_i^a dx + \omega^a \wedge \omega_i \quad . \tag{4.8}$$

Let θ be the Cartan form of L. Since L contains no derivatives, we have:

$$\theta = Ldx \quad .$$

Let θ' be the Cartan form of L'. Then,

$$\theta' = \theta + \lambda_a^i \theta_i^a$$

$$= Ldx + \lambda_a^i(z_i^a - h_i^a)dx + \lambda_a^i \omega^a \wedge \omega_i$$

$$= Ldx + (\lambda_a^i z_i^a - \lambda_a^i h_i^a)dx$$

$$+ \lambda_a^i(dz^a - z_j^a dx^j) \wedge \left(\frac{\partial}{\partial x^i} \lrcorner\ dx\right)$$

Now, let us take into account the constraints, i.e., set:

$$z_i^a = h_i^a .$$ (4.9)

Set

$$\theta'' = \theta'$$

restricted to the submanifold determined by conditions (4.9). The result is:

$$\theta'' = Ldx + \lambda_a^i(dz^a - h_j^a dx^j) \wedge \omega_i$$

$$= Ldx + \lambda_a^i dz^a \wedge \left(\frac{\partial}{\partial x^i} \lrcorner\; dx\right) - \lambda_a^i h_i^a dx$$ (4.10)

This is an especially simple formula. We can sum up as follows:

Theorem 4.1. Let a field-theoretic optimal control variational problem be defined by a Lagrangian function

$$L(x,z,u)$$

and by control functions

$$h_i^a$$

Let Λ be the space of variables

$$(\lambda_a^i)$$

Then, the geometric description of the extremals and their physical-geometric properties is governed by the Cartan form (4.9), an n-form on the manifold

$$X \times Z \times \Lambda \tag{4.11}$$

Remark. Here is a coordinate-free description of the manifold (4.11). Identify $M^1(X,Z)$ with the vector bundle over $X \times Z$, whose fiber above the point

$$(x,y) \; \varepsilon \; X \times Z$$

is the space

$$L(X_x, Z_z)$$

of linear maps $X_x \to Z_z$ between tangent spaces. Then, the Cartan form is a differential form on the vector bundle over $X \times Z$, whose fiber above the point (x,z) is

$$L(Z_z^d, X_x^d)$$

5.   THE DIFFERENTIAL EQUATIONS OF THE EXTREMALS OF
      OPTIMAL CONTROL VARIATIONAL PROBLEMS

Let us recapitulate. X is a manifold with coordinates (x^i). Z is a manifold with coordinates (z^a). Λ is a vector space with linear coordinates λ_a^i. U is a manifold.

L   is a map $X \times Z \times U \to R$

h_i^a are functions : $X \times Z \times U \to R$.

Set:

$$\theta \;=\; Ldx \;+\; \lambda_a^i \left(dz^a \wedge \left(\frac{\partial}{\partial x^i} \;\lrcorner\; dx \right) - h_i^a dx \right) \tag{5.1}$$

θ is an n-form on $X \times Z \times U \times \Lambda$,   called the <u>Cartan form</u> <u>of the variational problem</u>. (Notice that we have slightly changed notation from Section 4--the form denoted there by θ'' has become θ.)

    If

$$\phi: X \to Z \times U \times \Lambda$$

is a map, let

$$gr(\phi): X \to X \times Z \times U \times \Lambda$$

denote its <u>graph</u>.

<u>Definition</u>.   A map

$$\phi: X \to Z \times U \times \Lambda$$

is an <u>extremal</u> of the variational problem if the following conditions are satisfied:

$$(gr\ \phi)^*(A \;\lrcorner\; d\theta) \;=\; 0 \tag{5.2}$$

    for all $A \;\varepsilon\; V(X \times Z \times U \times \Lambda)$.

$$\phi^*(dz^a - h_i^a dx^i) \;=\; 0 \tag{5.3}$$

Let us now work out the explicit equations equivalent
to (5.2) in the local coordinates appearing in (5.1).

We can also write θ in the following form:

$$\theta = Ldx + \lambda_a^i \omega^a \wedge \omega_i \quad , \tag{5.4}$$

where

$$\omega^a = dz^a - h_j^a dx^j \tag{5.5}$$

$$\omega_i = \frac{\partial}{\partial x^i} \lrcorner dx \tag{5.6}$$

We also have

$$d\theta = dL \wedge dx + d\lambda_a^i \wedge (dz^a \wedge \omega_i - h_i^a dx)$$

$$\tag{5.7}$$

$$= \lambda_a^i dh_i^a \wedge dx$$

Suppose that the manifold U has coordinates

$$(u^\alpha) \quad .$$

Suppose ϕ is an extremal. Then,

$$(gr \ \phi)^* \left(\frac{\partial}{\partial u^\alpha} \lrcorner d\theta \right)$$

$$= (gr \ \phi)^* \left(\frac{\partial}{\partial u^\alpha} \lrcorner (dL \wedge dx + d\lambda_a^i \wedge \omega^a \wedge \omega_i + \lambda_a^i d\omega^a \wedge \omega_i) \right)$$

$$= (gr \ \phi)^* \left(\frac{\partial L}{\partial u^\alpha} dx + \lambda_a^i \left(\frac{\partial}{\partial u^\alpha} \lrcorner d\omega^a \right) \wedge \omega_i \right)$$

$$= (gr \ \phi)*\left(\frac{\partial L}{\partial u^{\alpha}} \ dx - \lambda^i_a \frac{\partial h^a_j}{\partial u^{\alpha}} \ dx^j \wedge \omega_i\right)$$

$$= (gr \ \phi)*\left(\frac{\partial L}{\partial u^{\alpha}} - \lambda^i_a \frac{\partial h^a_i}{\partial u^{\alpha}}\right) = 0 \qquad (5.7)$$

This calculation suggests that we investigate the relations

$$\frac{\partial}{\partial u^{\alpha}} (L - \lambda^i_a h^a_i) = 0 \qquad (5.8)$$

Set:

$$h = L - \lambda^i_a h^a_i \qquad (5.9)$$

Definition. The variational problem is said to be non-singular if the functions

$$\frac{\partial h}{\partial u^{\alpha}}$$

on $X \times Z \times U \times \Lambda$ are functionally independent, i.e., if their differentials are linearly independent.

Let us suppose, for simplicity, that the variational problem is non-singular. Set:

$$M = \text{submanifold of } X \times Z \times U \times \Lambda$$
$$\text{on which } \partial h/\partial u^{\alpha} = 0$$

We can then sum up relation (5.7) as follows:

<u>Theorem 5.1</u>. If $\phi: X \to Z \times U \times \Lambda$ is an extremal, then

$$(\text{gr } \phi)(X) \subset M \tag{5.10}$$

<u>Remark</u>. The functions x, λ, z <u>restricted to</u> M form a
coordinate system, since the problem is non-singular. In
classical language, this means that the control variables
u are "eliminated" by being expressed in terms of
(x, λ, z). This substitution is sometimes called the
<u>Legendre transform</u>.

Let us now suppose that

$$(x^i, z^a, \lambda^i_a)$$

are coordinates for M. We must work out the rest of the
conditions implied by (5.2). Suppose A is a vector
field on M. Then,

$$A \lrcorner \, d\theta = A \lrcorner \, (dL \wedge dx + d\lambda^i_a \wedge \omega^a \wedge \omega_i + \lambda^i_a d\omega^a \wedge \omega_i)$$

$$= A \lrcorner \, (dL \wedge dx) - d\lambda^i_a \omega^a(A) \wedge \omega_i$$

$$+ \lambda^i_a (A \lrcorner \, d\omega^a) \wedge \omega_i + \cdots \tag{5.11}$$

(The terms $\cdots$ denote differential forms in the ideal generated by ω^a, hence which are zero when pulled back under $(gr\ \phi)$.)

Suppose first, that A does not involve $\partial/\partial x$, i.e.,

$$A(x^i) = 0 \qquad\qquad (5.12)$$

Then,

$$\omega^a(A) = A(z^a) \qquad\qquad (5.13)$$

$$A \lrcorner d\omega^a = -A \lrcorner (dh_i^a \wedge dx^i)$$

$$= -A(h_i^a)dx^i \qquad\qquad (5.14)$$

Inset (5.13) and (5.14) into (5.11):

$$A \lrcorner d\theta = A(L)dx - A(z^a)d\lambda_a^i \wedge \omega_i$$

$$- A(h_j^a)dx^j \wedge \lambda_a^i \omega_i + \cdots$$

$$= A(L)dx - A(h_i^a)\lambda_a^i dx$$

$$- A(z^a)d\lambda_a^i \wedge \omega_i + \cdots \qquad\qquad (5.15)$$

Using (5.9),

$$A(L) - A(h_i^a)\lambda_a^i = A(h) + A(\lambda_a^i)h_i^a$$

Hence, (5.15) takes the following form:

$$A \lrcorner \, d\theta = (A(h) + A(\lambda_b^j)h_i^b)dx - A(z^b)d(\lambda_b^j\omega_j) + \cdots \quad (5.16)$$

Consider

$$(gr \; \phi)^*(\lambda_b^i\omega_i) \quad .$$

It is a form of degree (n-1) on X. Let us denote it as follows.

$$(gr \; \phi)^*(\lambda_b^i\omega_i) = \lambda_b^i(x)\omega_i \quad , \quad\quad (5.17)$$

where:

$$\lambda_b^i(x) = (gr \; \phi)^*(\lambda_b^i) \quad\quad (5.18)$$

We then have:

$$d(gr \; \phi)^*(\lambda_b^i\omega_i)) = \frac{\partial}{\partial x^i} (\lambda_b^i(x)) \quad . \quad\quad (5.19)$$

(Notice how the "divergence" differential operator appears very naturally at this stage!)

So far, A is any vector field tangent to M. Let us take it to be the vector fields corresponding to the coordinates (x,z,λ) for M:

Case 1: $A = \partial/\partial z^a$

Relations (5.2) and (5.16) then imply that:

$$\frac{\partial}{\partial x^i} (\lambda_a^i(x)) = \frac{\partial h}{\partial z^a} \quad\quad (5.20)$$

<u>Case 2</u>. $A = \partial/\partial\lambda_a^i$

Relations (5.2), (5.3) and (5.16) now imply that:

$$\frac{\partial}{\partial x^i} (z^a(x)) = -\frac{\partial h}{\partial(\lambda_a^i)} \qquad (5.21)$$

Equations (5.20) and (5.21) constitute partial differential equations which determine the extremal ϕ. They are in fact the <u>field-theoretic Hamilton equations</u>.

This whole process may be thought of as a generalization of the <u>Pontrjagin Principle</u> of optimal control theory.   In the next section I will recapitulate this material in a form that might be more useful for applications.

6.   THE MULTIPLE INTEGRAL VERSION OF THE PONTRJAGIN
     PRINCIPLE

In previous sections of the chapter, I have described the differential equations of a multiple-integral optimal control problem using the general differential-geometric formalism developed in my previous books, particularly LAQM, VB, and GPS.   Now, I will recast the formulas in a form that directly generalizes the usual Pontrjagin way of describing the simple-integral problems.   In this way,

the formulas can be used by those who do not understand
the elaborate differential geometric formalism I have
used to derive the equations.

Let us reformulate the underlying assumptions.
Suppose X,Z,U are finite dimensional vector spaces.

L   is a may $X \times Z \times U \to R$, called the Lagrangian.

f   is a map : $X \times Z \times U \to L(X,Z)$ called the
control law.

$\underline{z}$ denotes a map $X \to Z$

$\underline{u}$ denotes a map $X \to U$

A pair $(\underline{z},\underline{u})$ satisfies the control law if:

$$d\underline{z}(x) = \underline{f}(x, \underline{z}(x), \underline{u}(x)) \qquad (6.1)$$

Remark. $d\underline{z}(x)$ denotes the (Frêchét) differential of the
map z, which is, for fixed x, a linear map $X \to Z$.

The variational problem is now to extremize the
following action function,

$$\underset{\sim}{L}(\underline{z},\underline{u}) = \int_X L(\underline{z}(x), \underline{u}(x), x) \, dx \quad , \qquad (6.2)$$

over the set of pairs $(\underline{z},\underline{u})$ of maps which satisfy the
control law (6.1).

Let:

$$\Lambda = L(Z^d, X^d) \equiv \text{dual space to } L(X,Z) \quad . \qquad (6.3)$$

Let h' be the following real valued function $X \times Z \times U \times \Lambda$

$$h'(x,z,u,\lambda) = L(x,z,u) - \lambda(f(x,z,u)) \qquad (6.4)$$

Define a function

$$h: X \times Z \times \Lambda \to R$$

by the following rule:

$$h(x,z,\lambda) = \underset{u \epsilon U}{\text{extremum}} \, h'(x,z,\lambda,u) \qquad (6.5)$$

Remark. "Extremum" will typically mean maximum or minimum if h' is a concave or convex function of u.

Definition. h is the Hamiltonian of the problem. Formula (6.5) is the Pontrjagin method for defining h.

Now, introduce linear coordinates

$$(x^i)$$

for X, linear coordinates

$$(z^a)$$

for Z. Let

$$(\lambda^i_a)$$

be the corresponding linear coordinates for Λ. Thus,
h becomes a function

$$h(x^i, z^a, \lambda_b^j)$$

of these variables.

Definition. A pair $(\underline{z}, \underline{\lambda})$ of maps $X \to Z, \Lambda$, is said to
define an <u>extremal map for the variational problem in the</u>
<u>Hamilton-Pontrjagin sense</u> if the following differential
equations are satisfied:

$$\frac{\partial}{\partial x^i}\, (z^a(x)) \;=\; -\,\frac{\partial h}{\partial (\lambda_a^i)}\, (x, \, \underline{z}(x), \, \underline{\lambda}(x))$$

$$\text{(6.6)}$$

$$\frac{\partial}{\partial x^i}\, (\lambda_a^i(x)) \;=\; \frac{\partial h}{\partial z^a}\, (x, \, \underline{z}(x), \, \underline{\lambda}(x)) \;\;,$$

where:

$$z^a(x) \;=\; z^a(\underline{z}(x))$$

$$\lambda_a^i(x) \;=\; \lambda_a^i(\underline{\lambda}(x)) \;\;.$$

i.e., $(z^a(x), \, \lambda_a^i(x))$ are the linear coordinates of the
maps $\underline{z}, \underline{\lambda}$.

This Hamilton-Pontrjagin form of the calculus of
variations equations should be very useful in both physics
and engineering. First, notice that it has been formulated

in a way that makes sense for inequality constraints. To
my knowledge, no one has explicitly extended the now
classical Pontrjagin work on simple integral variational
problems with inequality constraints to multiple integral
problems. Second, notice that it gives a very convenient
form to the field equations of what the physicists call
classical fields. I believe that investigating the quantization of those equations, and their analogue in quantum
field theory, might prove useful in elementary particle
physics.

7.   THE FORMALISM APPLIED TO THE CLASSICAL VARIATIONAL
      PROBLEM

It is not yet clear how general an optimal control
problem can be handled with the formalism described in Section
4 to 6. To show its usefulness, I will now apply it to what
I call the "classical" problem.

Suppose X and Z are manifolds, with coordinates

$$(x^i, z^a) \quad .$$

Then, $M^1(X,Z)$ has coordinates

$$(x^i, z^a, z^a_i) \quad .$$

Let

$$L(x,z,\partial z)$$

be a function on $M^1(X,Z)$, called the _Lagrangian_. The
classical problem is to choose $x \to z(x)$ to extremize

$$\int L\left(x,z(x),\frac{\partial z}{\partial x}\right) dx \quad . \tag{7.1}$$

To convert it to one of form (4.3)-(4.4), introduce
control variables

$$u = (u_i^a) \quad . \tag{7.2}$$

L becomes a function

$$L(x,z,u) \tag{7.3}$$

of these variables.

Set

$$h_i^a(x,z,u) = u_i^a \tag{7.4}$$

Consider the problem of extremizing

$$\int L(x,\underline{z}(x),\underline{u}(x))\ dx$$

subject to the constraint

$$\frac{\partial z^a}{\partial x^i} = h_i^a$$

Using (7.4), it takes the form

$$L(x,\ \underline{z}(x),\ \underline{u}(x))\ dx \quad ,$$

subject to

$$\frac{\partial z^a}{\partial x^i} = u_i^a$$

This is obviously equivalent to the problem of extremizing (7.1).

Let us now compute the Hamiltonian, (5.9), for this special case. Let

$$\lambda_a^i$$

be the Lagrange multiplier varuables which are dual to the $h_i^a \equiv u_i^a$. Then, (5.9) takes the form

$$h = L(x,z,u) - \lambda_a^i u_i^a \qquad (7.5)$$

Follow the ideas of Section 5. Restrict the variables (x,z,u,λ) be the following relation:

$$\frac{\partial h}{\partial u_i^a} = 0 \qquad (7.6)$$

With h given by (7.5), Equations (7.6) take the following form:

$$\lambda_a^i = \frac{\partial L}{\partial u_i^a} \qquad (7.7)$$

We can also write down Equations (5.20)-(5.21):

$$\frac{\partial}{\partial x^i} (\lambda^i_a(x)) = \frac{\partial h}{\partial z^a} = \frac{\partial L}{\partial z^a} \tag{7.8}$$

$$\frac{\partial}{\partial x^i} (z^a(x)) = - \frac{\partial h}{\partial (\lambda^i_a)} = u^a_i \tag{7.9}$$

Equations (7.7)-(7.9) are then <u>equivalent</u> to the classical Euler-Lagrange equations for the unconstrained multiple integral variational problem, (7.1). Equation (7.9) shows that $u^a_i(x)$ are the derivatives of $z^a(x)$. (7.7) and (7.8) <u>together</u> give:

$$\frac{\partial}{\partial x^i} \left(\frac{\partial L}{\partial u^a_i} (x, z(x), \partial z(x)) \right) = \frac{\partial L}{\partial z^a} \tag{7.10}$$

which is, of course, just the classical Euler-Lagrange equation.

I leave this topic now, planning to return to a systematic geometric study of distributed parameter optimal control variational problems.

Chapter 6

A GENERAL FORM OF CIRCUIT THEORY

1. INTRODUCTION

Continuing (from Volume IX) to develop circuit theory as
a prototype mathematical theory, in this chapter I discuss
Tellegen's Theory, the Brayton-Moser [1] form of the circuit
theory equations, and a general differential-geometric setting
for the ideas. I hope to clear the ground for further
developments in circuit theory itself, for generalizations
to other areas (such as biology, chemistry, economics),
and, perhaps, suggest new areas for mathematical research!

2. TELLEGEN'S THEOREM

Physically, Tellegen's Theorem says that the total
power in a circuit is zero. Since "energy" is the derivative of power, this is just, of course, another manifestation
of "conservation of energy". Mathematically, its proof is
(now) an easy consequence of the fact that the two Kirchoff

circuit laws are dual to each other.

Here is the standard set-up which leads to Tellegen's Theorem. (See Desoer and Kuh [1], Anderson and Vongpanitlerd [1], Brayton and Moser [1], Smale [1] and Volumes II and IX background and references). Let (B, A, φ) be an oriented graph, with:

> B = set of branches
>
> A = set of nodes
>
> φ: B $\to$ A $\times$ A the incidence map.

Let:

> I = vector space of formal linear
>   combination of branches with real
>   coefficients $\equiv$ space of <u>currents</u>.
>
> V = dual space of I
> $\equiv$ space of <u>voltages</u>.

An element i $\in$ I thus can be written as:

$$i = \sum_{b \in B} i_b b. \tag{2.1}$$

The real number i_b is called the current <u>through the branch b</u>. We shall consider that:

> For each branch b $\in$ B, i_b is a
> real valued function on I.

The incidence map $\varphi: B \to A \times A$ defines a <u>boundary map</u>:

$$\beta(i) = \sum_{b \in B} i_b(a_1 - a_2),$$

where:

$$\varphi(b) = (a_1, a_2) \in A \times A.$$

The <u>Kirchhoff current law</u> subspace of I, abbreviated to

KCL,

is the linear subspace of I consisting of the currents i
such that:

$$\beta(i) = 0.$$

The <u>Kirchoff voltage law</u> subspace of V, denoted by

KVL,

is the orthogonal complement of KCL in V, i.e. the set of
$v \in V(\equiv I^d)$ such that

$$v(KCL) = 0. \qquad\qquad (2.2)$$

Each branch $b \in B$ defines a real-valued function

$$v_b: V \to R$$

as follows: For i of form (2.1),

$$v_b(i) = i_b$$

v_b is called the voltage <u>through</u> the <u>branch b</u>.

Remark: v_b is a element of V^d, which is equal to I. Similiarly, i_b is a element of I^d = V. v_b and i_b are really
just _dual bases_, in the linear algebra sense, of the vector
spaces I and V.

Now, for each $v \in V$, let the _value_ of the function v_b
on v be _also_ denoted by v_b. This notation is not strictly
correct but is the usual one in the engineering literature.
Similiarly, for $i \in I$, let i_b _also_ denote the value of the
function i_b on i. With these conventions, we can write a
voltage $v \in V$ and current $i \in I$ in the following forms:

$$v = \sum_{b \in B} v_b\, b \tag{2.3}$$

$$i = \sum_{b \in B} i_b\, b. \tag{2.4}$$

This convention has the property that:

$$v(i) = \sum_{b \in B} v_b i_b, \tag{2.5}$$

where $i \to v(i)$ is the linear form v defines on I. (Recall
that V is _defined_ as the dual space of I.)

Remark. The precisely correct version of the basic identity
(2.5) is:

$$v(i) = \sum_{b \in B} v_b(v) i_b(i). \tag{2.6}$$

Definition. If $(v, i) \in V \times I$ are a pair of voltages and currents, then the real number

        v(i)

is called the <u>power</u> in the voltage-current pair (v, i), and
is denoted by:

        P(v, i).

<u>Theorem 2.1</u>. (Tellegin's Theorem)

    If v ∈ KVL, ⊂ V, i ∈ K ⊂ L ⊂ I, then

        P(v, i) = 0.   (2.7)

    The <u>proof</u> of (2.7) is now a <u>tautology</u> - since we have
<u>defined</u> KVL as the orthogonal complement of KCL in V under
the bilinear pairing defined by P(,). It is a standard
<u>exercise</u> in linear algebra to show that this is equivalent
to KVL in its usual form, i.e. that "voltage drop" around
a closed loop is zero.

3. THE BRAYTON-MOSER THEOREM

    So far, there has been no "dynamics." To put this in,
we set:

        M = V × I.

Consider M as a manifold, with i_b, v_b, the current and
voltage in each branch b, considered as real-valued functions
on M. Consider t as a time parameter,

$$-\infty < t < \infty.$$

"Dynamics" defines curves

$$t \rightarrow (i(t), v(t))$$

in M. The differential equations which define these curves
are given by the constitutive relations in each of the
branches b.

What we shall call the Brayton-Moser Theorem is, in a
sense, the dynamic version of Tellegen's Theorem, i.e.
basically a consequence of the "duality" between voltages
and currents. I shall more-or-less follow Smale's exposition [1] of the Brayton-Moser result.

To define the constitutive relations in the most useful differential geometric way, for each branch b $\in$ B construct a two dimensional manifold

$$M_b,$$

with i_b, v_b, as coordinates on M_b. The constitutive
relation of the branch b will be set of differential equations for curves in M_b. There are three types:

Resistive constitutive relation:

A functional (i.e. zero-th order differential) relation between i_b, v_b, say:

$$f_b(v_b, i_b) = 0 \qquad (3.1)$$

Capacitive relation

A differential equation of the following form:

$$i_b = f_b(v) \frac{dv_b}{dt}. \qquad (3.2)$$

Inductive relation

A differential equation of the following form:

$$v_b = f_b(i) \frac{di_b}{dt}. \qquad (3.3)$$

Recall that:

$$M = V \times I,$$

i_b, v_b are functions on M. Set:

$$\theta = \sum_b v_b di_b, \qquad (3.4)$$

a one-form on M.

Remark: M can also be identified with the cotangent bundle
to the vector space I. With this identification, (3.4)
shows that θ agrees with the 1-form defining the usual
contact structure on the cotangent bundle. See GPS. This

"coincidence" (which is really not, of course) is very
significant for possibilities of generalization.

Theorem 3.1. The submanifold

$$(KCL) \cap (KVL) \subset M \equiv I \times V$$

is a maximal dimension integral submanifold of θ, i.e. forms
what is called a Lagrangian submanifold of θ.

Exercise. Prove Theorem 3.1. It follows from the same
underlying algebraic fact as Tellegen's Theorem.

   Now, the branches fall into three mutually exclusive
types, denoted by

$$B_R, \ B_C, \ B_{IN},$$

called the resistive, capacitive, inductive branches, depending on which of the three types of constitutive
relations described above holds. Set:

$$\theta_R = \sum_{b \in B_R} v_b di_b \qquad (3.5)$$

$$\theta_C = \sum_{b \in B_C} v_b di_b \qquad (3.6)$$

$$\theta_{IN} = \sum_{b \in B_{IN}} v_b di_b. \qquad (3.7)$$

Since

$$B = B_R \cup B_{CU} B_{IN},$$

we have:

$$\theta = \theta_R + \theta_C + \theta_{IN}. \tag{3.8}$$

To prove the βrayton-Moser Theorem we must discuss separately the differential geometric properties of the differential forms θ_R, θ_C, θ_{IN}.

Resistors. Set

$$M' = \{(v, i) \in M: f_b(v_b, i_b) = 0 \tag{3.9}$$

$$\text{for all } b \in B_R\}.$$

In other words, M' consists of all voltage-current pairs whose resistive branch voltages and currents satisfy the resistive constitutive relations. (Since they are "functional", but not "differential", conditions this makes sense. This is the reason why they are treated differently from the capacitive and inductive branches, at least in the Brayton-Moser formulation. Of course, from the "Lagrange-Rayleigh Port" point of view described in Volume IX they should all be treated on a more equal footing).

Further, this submanifold M' has the following property:

$$d\theta_R = 0 \text{ on } M'. \tag{3.10}$$

This is the characteristic <u>reciprocal constitutive relation</u>.
In this case, it is a trivial consequence of the fact that
a 2-form is zero on a 1-dimensional submanifold (or) that
iterated <u>ordinary</u> derivatives commute.) In the obvious
n-port generalization of this, (3.9) must be <u>assumed</u>.
Brayton has extensively discussed this [1].

In order to discuss the capacitors and indicators in
the Brayton-Moser formulation, we must discuss the auxillary
concept of a "gradient curve associated with a Riemannian
metric."

<u>Definition</u>. Let N be a manifold,

$$\gamma: V(N) \times V(N) \to F(N)$$

a symmetric F(N)-bilinear form on tangent vector fields.
(Thus, if γ is in addition algebraically <u>non-degenerate</u> it
defines a <u>Riemannian metric structure on N</u>.) A curve
$t \to \sigma(t)$ N is a <u>gradient curve</u> with respect to γ if there
exists a 1-form η on N such that:

$$\gamma(\sigma'(t), v) = \eta(v) \tag{3.11}$$
$$\text{for all } v \in N_{\sigma(t)}, \text{ all } t,$$

<u>Remark</u>: Using the standard "contraction" symbol $\lrcorner$, we can

write (3.11) as:

$$\eta = \sigma'(t) \; \lrcorner \; \gamma.$$

If further

$$d\eta = 0, \qquad\qquad\qquad (3.12)$$

then the gradient curve is said to be <u>integrable</u>.

<u>Remark</u>: Here is what is meant in local coordinates, say
(x_j), $1 \leq j$, $k \leq n = \dim N$. Say that:

$$\gamma = g_{jk} dx_j \cdot dx_{ik} \qquad\qquad\qquad (3.13)$$

(The "product" on (3.13) is the symmetric product of
differential forms.) Condition (3.12) means that (locally)
there is a function $fx \rightarrow f(x)$ such that:

$$\eta = df = \frac{\partial f}{\partial x_j} \, dx_j.$$

Let $t \rightarrow (x_j(t))$ be the curve in R^n which represents the
curve $t \rightarrow \sigma(t)$ in these local coordinates. Then, condition
(3.11) means that $t \rightarrow x(t)$ satisfies the following set of
ordinary differential equations:

$$2g_{jk}(x(t)) \frac{dx_k}{dt} = \frac{\partial f}{\partial x_j} (x(t)). \qquad\qquad (3.14)$$

Of course, if the form η is non-singular, i.e. if

$$\det(g_{jk}) \neq 0,$$

then equations (3.14) can be solved for $\frac{dx_j}{dt}$. In this form, one will recognize the usual definition of "gradient curve" of Riemannian geometry and Morse Theory. (See Milnor [1]). If η is a skew-symmetric form, then the equations (3.14) can be written in Hamiltonian form, with "f" the Hamiltonian Thus, the Brayton-Moser equations (which we shall describe in a moment as a certain set of gradient curve equations) stand in the same relation to electric circuit theory as the Hamilton equations stand to classical mechanics.

Return to circuit theory on a graph (B, A, φ). Let $b \in B$ be a given branch:

Let $\theta_b = v_b di_b$. The constitutive relation defines a family of curves in the space of variables

$$(v_b, i_b),$$

which satisfy a given type of differential equations. We shall examine the inductive and capacitive type, where the differential equations are given by (3.3) and (3.2). In each case, we shall show that the differential equations imply that the curves which satisfy them can be described by relations of the form (3.11), with θ_b the form η and with a certain symmetric differential form γ_b on R^2 called the Brayton-Moser form.

Inductance branches

The standard circuit theory symbol for such a branch
is:

The differential equation relating branch current and
voltage

$$(i_b, v_b),$$

is given by (3.3), i.e.

$$v_b(t) = f_b(i) \frac{di_b}{dt} .$$

along such a curve in (i_b, v_b)-space,

$$\theta_b = v_b di_b = f_b(i) \frac{di_b}{dt} di_b$$

$$= \sigma'_b(t) \lrcorner \gamma_b, \tag{3.15}$$

where $\sigma_b(t)$ is the curve $t \to (i_b(t), v_b(t))$ in branch
current-voltage space, $t \to \sigma'_b(t)$ is its tangent vector
curve, and γ_b is the following symmetric quadratic differential form:

$$\gamma_b = 2f_b di_b \cdot di_b. \tag{3.16}$$

Formula (3.15) is the Brayton-Moser formula for inductive
branches. γ_b is called the inductive Brayton-Moser metric.

It exhibits the current-voltage curves in the inductive
branches as (non-integrable) gradient curves with respect

to the 3-M metric (3.16).

Capacitive branches

The standard circuit theory symbol is:

The differential equation relating branch currents and voltage is:

$$i_b(t) = f_b(v) \frac{dv_b}{dt}. \tag{3.17}$$

Consider:

$$\theta_b = v_b di_b.$$

It can also be written as follows:

$$\theta_b = d(v_b i_b) - i_b dv_b \tag{3.18}$$

$$= d(P_b) - \theta'_b, \tag{3.19}$$

where

$$P_b = v_b i_b \equiv \text{the } \underline{\text{power}} \text{ in the branch b.}$$

$$\theta'_b = i_b dv_b. \tag{3.20}$$

Remark: What we have done here is to apply a Legendre transform, familiar from Thermodynamics (see GPS, Chapter 6), to interchange the role of current and voltage.

Now, we can combine (3.17) and (3.20) in the same way as
we handled the inductive branch. Along a solution of (3.17),

$$\theta_b' = f_b \frac{dv_b}{dt} \, dv_b$$

$$= \sigma_b'(t) \, \lrcorner \, \gamma_b, \qquad (3.21)$$

where:

$$\gamma_b = 2f_b dv_b dv_b \qquad (3.22)$$

γ_b is called the capacitive <u>Brayton-Moser metric</u>. Combining (3.21) and (3.19), we see that we can write the differential equations of the capacitive constitutive relations
in the following coordinate-free form:

$$\theta_b - dP_b = - \sigma_b'(t) \, \lrcorner \, \gamma_b. \qquad (3.23)$$

<u>Remark</u>: Compare (3.15) and (3.23). Note the difference
in sign!

We can now put all of this together:

<u>Theorem 3.2.</u>   (Brayton-Moser)

Let (B, A, φ) be an oriented graph, with I and V the
vector space of currents and voltages on the graph. Let

$$KCL \subset I$$

$$KVL \subset V$$

be the linear subspaces determined by the Kirchoff laws.

Suppose given a set of constitutive relations, determining a decomposition of B into three subsets, the resistors, capacitors, inductors. Let

$$M' \subset I \times V$$

denote the subset of $(i, v) \in I \times V$ such that:

> $i \in KCL$
>
> $v \in KVL$.
>
> For the resistive branches b, (i_b, v_b) satisfies the resistive constitutive relations.

Suppose that M' is a manifold.

With these hypotheses, there is a symmetric, F(M')-bilinear mapping

$$\gamma: V(M') \times V(M') \to F(M')$$

and a 1-form $\eta \in F^1(M')$ such that:

> $\alpha\eta = 0$
>
> Every curve $t \to \sigma(t)$ in $I \times V$ which satisfies the Kirchoff laws and the differential equations in each of the branches satisfies the following differential equation:
>
> $$\sigma'(t) \lrcorner \gamma = \eta. \qquad\qquad (3.24)$$

Proof. We have seen that a metric can be defined at
each capacitive or inductive branch. Summing up these
relations, i.e. (3.15) and (3.23), together with the re-
ciprocity of the resistive constitutive relation, Theorem
3.1, and relations (3.5)-(3.8), gives (3.24).

Remark: Under suitable assumptions (roughly, the same
conditions that the capacitor voltages and inductor currents
be choosable as state variables) one can show that γ is a
non-degenerate form, hence defines a Riemannian metric for
M'. See Smale, [1]. Unfortunately, this metric is of
indefinite sign, and not very much is known about global
properties of gradient curves of such metrics, so that the
metric does not seem to be of much help for proving theorems
about networks. Presumably, what we need is a simultaneous
improvement in the state of the art in studying properties
of such gradient curves. (This would also be important for
other areas of mathematics and physics, e.g. the theory of
partial differential equations and relativity.)

Remark: The metric γ - resulting from summing up the metrics
of each of the capacitive and inductive branches - is of the
following form:

$$\gamma = \sum_{\text{inductors}} f_b(i_b)di_b \cdot di_b$$

$$\text{(3.25)}$$

$$- \sum_{\text{capacitors}} f_b(v_b)dv_b \cdot dv_b.$$

We can now abstract from these ideas a general scheme for introducing differential equations, interactions, and interconnections based on a graph.

4. DYNAMIC INTERCONNECTION STRUCTURES

Let (B, A, φ) be a directed graph, with branches B, nodes A, incidence map $\varphi: B \to A \times A$. Suppose given, for each $b \in B$, a manifold

$$M_b,$$

and the following geometric structures on M_b:

a 1-form θ_b (4.1)

An $F(M_b)$-bilinear form

$$\gamma_b: V(M_b') \times V(M_b) \to F(M_b).$$ (4.2)

(θ_b, γ_b) may be said to define a <u>constitutive relation</u> at the branch b.

<u>Remark</u>: Notice that we have not specified whether γ is

symmetric or skew-symmetric. As we have seen, the former
assumption is appropriate for circuit theory, the latter
for classical mechanics.

Form:

> M = direct product of the manifolds
> M_b, as b runs over the branches of   (4.3)
> the graph.

$$\theta = \sum_{b \in B} \theta_b$$
$$\gamma = \sum_{b \in B} \gamma_b . \qquad (4.4)$$

Definition. An interconnection of the elements represented by the branches B is a submanifold

> $M' \subset M$

on which $d\theta = 0$.

Given such a submanifold M', set:

> $\theta' = \theta$ restricted to M'
> $\gamma' = \gamma$ restricted to M'

We can form the gradient curves of θ' with respect to γ'.
They are - just as in the circuit theory case - to be considered as the dynamical equations.

In summary, we see that the Brayton-Moser description
of the electric circuit equations is very well suited to

an elegant geometric generalization. In fact, the mathematics is very close to that of thermodynamics (See GPS,
Chapter 6). Of course, Meisner points out that circuit
theory is really an example of a thermodynamic system [1].
This idea has been further developed and generalized (e.g.
to classical systems) by Katchalsky, Oster and Perel son
[1].

Chapter 7

THE EQUATIONS OF MOTION OF MECHANICAL SYSTEMS

1.   INTRODUCTION

As Gantmacher points out in the preface to his "Lectures
to analytical mechanics", the essence of the subject is the
systematic deduction of the laws of motion of mechanical systems
from general mathematical principles. In this book, I have
decided to emphasize Lagrange's equations, and particularly,
the differential geometric way of looking at them pioneered
by E. Cartan in "Leçons sur les invariants intégraux". (In
fact, one great virtue of Cartan's viewpoint is that the
"Lagrangian" and "Hamilton-Jacobi" viewpoints are much more
unified than in the classical theory, since they basically
only differ by a choice of coordinates. In my "Geometry,
physics and systems" I have described how the most general
equations of mechanics may be described using Cartan's method,
and this chapter is basically an elaboration of the work
presented there.)

2.   LAGRANGE'S EQUATIONS IN LOCAL COORDINATES

Let n be an integer. (It will be identified with
the number of degrees of freedom of the mechanical system.)

Let

$$q = (q^1, \ldots, q^n)$$

be a set of real variables, called the <u>position coordinates</u>.
Introduce another set of n-vectors

$$\dot{q} = (\dot{q}^1, \ldots, \dot{q}^n)$$

called the <u>velocity variables</u>. Let

$$t$$

be a real variable, called the <u>time</u>, varying over an interval
T of real numbers. Choose the following range of indices and
the summation convention on these indices:

$$1 \le i,j \le n \quad .$$

Let

$$L(q, \dot{q}, t)$$

be a function of these variables. It is called the <u>Lagrangian</u>

Let $F_i(q, \dot{q}, t)$ be an n-type of functions of the
position-velocity-time variables. They are called the <u>compon-
ents of the forces</u>.

Consider now the following set of second-order ordinar
differential equations, to be solved for a curve

$$t \to q(t) \equiv (q^i(t))$$

in configuration space.

$$\frac{d}{dt}\left(\frac{\partial L}{\partial \dot{q}^i}\left(q(t),\frac{dq}{dt},t\right)\right) - \frac{\partial L}{\partial q^i}\left(q(t),\frac{dq}{dt},t\right) = F_i\left(q(t),\frac{dq}{dt},t\right)$$

$$(2.3)$$

They are called <u>Lagrange's equations</u>.

3.   NEWTON'S LAWS OF MOTION

Specialize L to be the following form:

$$L = \frac{1}{2} m_{ij}\dot{q}^i\dot{q}^j \tag{3.1}$$

The coefficients (m_{ij}) are constants. (Typically, they are determined by the <u>masses</u>.)

Lagrange's equations (2.3) take the form:

$$m_{ij}\frac{d^2q^j}{dt^2} = F_i\left(q(t),\frac{dq}{dt},t\right) \tag{3.2}$$

This is basically Newton's Law:

$$\text{Mass} \times \text{Acceleration} = \text{Force} \tag{3.3}$$

Systems of particles moving in space can readily be described by equations of form (3.3). Let

$$\vec{x} = (x,y,z)$$

denote a vector of R^3. Suppose the system consists of N particles. Let

$$\vec{x}_1, \ldots, \vec{x}_N$$

denote their position vectors. Let

$$m_1, \ldots, m_N$$

denote their masses.

In Newtonian mechanics one postulates that the forces $\vec{F}_1, \ldots, \vec{F}_N$ depend on the positions and velocities of the particles. Newton's Law then takes the form:

$$
\begin{aligned}
m_1 \frac{d^2\vec{x}_1}{dt^2} &= \vec{F}_1\!\left(\vec{x}_1, \frac{d\vec{x}_1}{dt}, \ldots, \frac{d\vec{x}_N}{dt}, t\right) \\
&\;\;\vdots \\
m_N \frac{d^2\vec{x}_N}{dt^2} &= \vec{F}_N(\vec{x}_1, \ldots, t)
\end{aligned}
\tag{3.4}
$$

These equations can obviously be put into form (3.2). For example, set:

$$n = 3N$$

$$q = (x_1, y_1, z_1, x_2, \ldots, z_N) \quad .$$

The mass matrix (m_{ij}) is then a diagonal matrix of the form:

$$
\begin{pmatrix}
m_1 & & & & & \\
& m_1 & & & 0 & \\
& & m_1 & & & \\
& & & m_2 & & \\
& 0 & & & \ddots & \\
& & & & & m_N
\end{pmatrix}
$$

The __kinetic energy__ of the system of particles is

$$\frac{1}{2} m_1 |\vec{\dot{x}}_1|^2 + \cdots + \frac{1}{2} m_N |\vec{\dot{x}}_N|^2$$

Notice that this just equals the function L, as defined by (3.1). This identification is especially important for setting up Lagrange's equations in complicated practical situations. (See Whittaker [1], Pars [1].)

Thus, we see that the Lagrange and Newton form of the equations of motion only differ by a trivial change in notation. However, there is a fundamental differential-geometric difference between them. As we shall see, the Lagrange equations are much more "covariant" than Newton's, since they __make sense__ for arbitrary manifolds, while Newton's seem rather closely tied to vector spaces.

4.   THE LAGRANGE EQUATIONS IN COORDINATE FREE FORM

Let q^i, $\dot{q}^i$, t, L, F_i be as in Section 2. Set:

$$\theta = Ldt + \frac{\partial L}{\partial \dot{q}^i} (dq^i - \dot{q}^i dt) \tag{4.1}$$

θ is a one-differential form, in the variables $(q, \dot{q}, t)$. It is called the __Cartan form__, corresponding to the Lagrangian L. Set:

$$\theta^i = dq^i - \dot{q}^i dt \qquad (4.2)$$

$$\omega_i = d\left(\frac{\partial L}{\partial \dot{q}^i}\right) - \frac{\partial L}{\partial q^i} dt \qquad (4.3)$$

Let us now compute $d\theta$, using $(4.1)-(4.3)$:

$$d\theta = \frac{\partial L}{\partial \dot{q}^i} d\dot{q}_i \wedge dt + \frac{\partial L}{\partial \dot{q}^i} dq^i \wedge dt + d\left(\frac{\partial L}{\partial \dot{q}^i}\right) \wedge (dq^i - \dot{q}^i dt)$$

$$- \frac{\partial L}{\partial \dot{q}^i} d\dot{q}^i \wedge dt$$

Also,

$$\theta^i \wedge \omega_i = (dq^i - \dot{q}^i dt) \wedge \left(d\left(\frac{\partial L}{\partial \dot{q}^i}\right) - \frac{\partial L}{\partial q^i} dt \right)$$

$$= dq^i \wedge d\left(\frac{\partial L}{\partial \dot{q}^i}\right) - \frac{\partial L}{\partial q^i} dq^i \wedge dt - \dot{q}^i dt \wedge d\left(\frac{\partial L}{\partial \dot{q}^i}\right)$$

We see that:

$$d\theta = \omega_i \wedge \theta^i \qquad (4.4)$$

Remark. From my point of view, this is a <u>fundamental formula</u> <u>of analytical mechanics</u>. I believe that it was first given in IM, Vol. X, where it was used to study the Euler-Lagrange

variational equations in the case where the Lagrangian L
was degenerate. As we shall now show, it is also a basic
formula for the traditional problems of analytical mechanics.

We can now put the Lagrange equations (2.3) into a
coordinate-free form. Let

$$t \rightarrow \sigma(t)$$

be a curve in configuration space. In terms of the configuration coordinates (q^i), let

$$q(t) \equiv (q^i(\sigma(t))$$

be the coordinates of this curve.

Let $t \rightarrow \sigma_1(t)$ be the <u>prolongation</u> of σ, to be a
curve in $(q, \dot{q}, t)$-space, as follows:

$$\dot{q}^i(\sigma_1(t)) = \frac{d}{dt} q^i(t) \quad . \tag{4.5}$$

Let $t \rightarrow \sigma_1'(t)$ be the tangent vector curve to σ_1. Then,
we have:

$$\theta^i(\sigma_1'(t)) = 0 \tag{4.6}$$

Combine (4.4) and (4.6)

$$\sigma_1'(t) \,\lrcorner\, d\theta = \omega_i'(\sigma_1'(t))\theta^i \tag{4.7}$$

Now,

$$\omega_i(\sigma_1'(t)) \;=\; d\!\left(\frac{\partial L}{\partial \dot{q}^i}\right)(\sigma_1'(t)) \;-\; \frac{\partial L}{\partial q^i}(\sigma_1'(t))$$

$$(4.8)$$

$$=\; \frac{d}{dt}\!\left(\frac{\partial L}{\partial \dot{q}^i}\!\left(q(t),\, \frac{dq}{dt},\, t\right)\right) \;-\; \frac{\partial L}{\partial q^i}\!\left(q(t),\, \frac{dq}{dt},\, t\right)$$

Set:

$$\underline{F} \;=\; F_i(dq^i - \dot{q}^i dt) \qquad\qquad (4.9)$$

$\underline{F}$ is called the <u>force form</u>. Combining (4.7) and (4.8), we see that we have proved the following:

<u>Theorem 4.1.</u> A curve $t \to q(t) \equiv \sigma(t)$ in configuration space is a solution of the Lagrange equations (2.3) if and only if:

$$\sigma_1'(t) \;\lrcorner\; d\theta \;=\; \underline{F}(\sigma_1'(t)) \qquad\qquad (4.10)$$

for all t.

(4.10) is the basic formula, giving the Lagrange equations in coordinate-free form. Here is another useful geometric way of looking at these equations.

Let X be the manifold R^{2n+1}, with variables $(q, \dot{q}, t)$. Consider a vector field A on X such that:

$$A \;\lrcorner\; d\theta \;=\; \underline{F} \qquad\qquad (4.11)$$

$$\theta^i(A) \;=\; 0 \qquad\qquad (4.12)$$

$$A(t) \;=\; 1 \qquad . \qquad\qquad (4.13)$$

The solutions of the Lagrange equations (4.11)-(4.13) are the orbit curves of A.

5.   THE CONFIGURATION SPACE AS A MANIFOLD

So far, we have been working in the classical way, in terms of specific coordinates used to describe the position (or "configuration") of the mechanical system. Let us now reformulate the basic concepts in terms of manifold theory.

Let Q be a manifold. $T(Q)$ denotes its tangent vector bundle. T denotes an interval of real numbers, parameterized by the real "time" variable t. Set:

$$X = T(Q) \times T .$$

To identify with previous work, let

$$n = \dim Q$$

(q^i) be a coordinate system for Q.

A coordinate system $(q^i, \dot{q}^i)$ for $T(Q)$ can be defined as follows:

$$(q^i(v)) = q^i(q_0)$$

$$\text{for } v \in Q_{q_0}, \quad q_0 \in Q .$$

$$(\dot{q}^i(v) = dq^i(v)$$

$$\text{for } v \in T(Q) .$$

These coordinates are called <u>Newtonian coordinates</u> for $T(Q)$. Together with the coordinates t for T, we obtain a coordinate system

$$(q, \dot{q}, t)$$

for $T(Q) \times T = X$.

The manifold X is the underlying manifold on which most objects in mechanics are defined.

> The Lagrangian L is a real-valued function on X. The Cartan form θ is a one-form on X. (The assignment $L \to \theta$ is "functorial", i.e., independent of the coordinates used to defined it.)
>
> The force-law form
>
> $$\underline{F}$$
>
> is a one-form on X. It is in the Pfaffian system generated by the contact forms θ^i.

X can be readily identified with a <u>mapping element</u> space, namely:

$$X = M^1(T, Q) ,$$

the first-order mapping elements (or "jets", in the terminology of Ehresmann) of maps $T \to Q$. This identification is especially useful in understanding how the formalism may be generalized to cover mechanical systems with an infinite number of degrees of freedom, e.g., fields.

The manifold θ is called the configuration space manifold. We can now sum up these ideas in the following way:

Definition. A mechanical system (without constraints) is defined by giving a triple

$$(Q, L, \underline{F})$$

where:

Q is a manifold,

L: $T(Q) \times T \to R$ is a Lagrangian function,

$\underline{F}$ is a one-form on $T(Q) \times T$ which, in local coordinates, is a linear combination of the contact forms θ^i.

EXAMPLES OF MECHANICAL SYSTEMS

a)   A single particle moving in R^3

Denote a vector of R^3 by $\vec{x}$. In the usual way of writing three-vectors,

$$\vec{x} = (x, y, z) \ .$$

Consider a particle of mass m. Newton's Law takes the form:

$$m \frac{d^2 \vec{x}}{dt^2} = \vec{F}\left(\vec{x}, \frac{d\vec{x}}{dt}, t\right) \quad .$$

Suppose

$$\vec{F} = (F_x, F_y, F_z) \quad .$$

Then

$$Q = R^3 \quad .$$

$T(Q)$ is the space of variables

$$(x, \vec{\dot{x}}) \equiv (x, y, z, \dot{x}, \dot{y}, \dot{z})$$

$$L = \frac{1}{2} m |\vec{\dot{x}}|^2 = \frac{1}{2} m (\dot{x}^2 + \dot{y}^2 + \dot{z}^2)$$

$$\underline{F} = f_x \, dx + F_y \, dy + F_z \, dz \quad .$$

b) <u>A particle constrained to move on a curve of surface
    in R^3</u>

For example, constrain the particle to a surface of R^3. Suppose the coordinates on the surface are

$$(q, q_2) \quad .$$

Then, the surface is defined (locally) by vector-valued functions

$$\vec{x}(q_1, q_2) \quad .$$

The configuration manifold is the surface itself. Here is how one constructs L and $\underline{F}$. (Why this is the correct choice will be explained later on.)

$$L(q,\dot{q}) \;=\; \frac{1}{2}\, m\!\left(\frac{\partial \vec{x}}{\partial q_1}\,\dot{q}_1 \;+\; \frac{\partial \vec{x}}{\partial q_2}\,\dot{q}_2\right)^2$$

$$\underline{F} \;=\; F_x(x(q),\, y(q),\, z(q))\; d(x(q)) \;+\; \cdots$$

(In other words, $\underline{F}$ = the force differential form <u>restricted to the surface</u>.)

Note that L is essentially the <u>Riemannian metric induced on the surface</u> from the flat metric

$$dx^2 + dy^2 + dz^2$$

on R^3. In the case where the forces are zero, the solutions of Lagrange's equations are (after a change in parameterization) just the <u>geodesics</u> of the surface. Thus, in this case, "mechanics" essentially reduces to "geometry".

c)   <u>A rigid body</u>

Consider a set of particles in R^3 which are constrained to move "rigidly", i.e., so that the distances between them are constant. The key feature of this example is that Q can be identified with the underlying manifold of a Lie group, namely, the <u>group of rigid motions of</u> R^3, i.e., the <u>automorphism group of Euclidean geometry</u>. (It is for this reason that this group is often called the <u>Euclidean group</u>.)

Suppose the rigid body consists of N particles, whose positions are:

$$\vec{x}_1, \ldots, \vec{x}_N$$

The "rigidity" condition means that:

$$|\vec{x}_i - \vec{x}_j| = \text{constant} \equiv c_{ij} \qquad (5.1)$$

$$\text{for} \quad 1 \leq i.j \leq n \quad .$$

Let Q be the subset of

$$R^3 \times \ldots \times R^3 = R^{3N}$$

satisfying condition (4.10).

A <u>rigid motion</u> of R^3 is a diffeomorphism

$$g: R^3 \to R^3$$

such that:

$$|g(\vec{x}) - g(\vec{y})| = |\vec{x} - \vec{y}|$$

$$\text{for} \quad \vec{x}, \vec{y} \in R^3 \quad .$$

They form a group, denoted as G. G is a 6-<u>parameter Lie</u> <u>group</u>.

Consider G acting on R^{3N} as follows:

$$g(\vec{x}_1, \ldots, \vec{x}_N) = (g\vec{x}_1, \ldots, g\vec{x}_N) \qquad (5.2)$$

Note that G, acting this way, <u>preserves the relations</u> (4.10), i.e., acts on the submanifold Q of R^{3N}.

<u>Definition</u>. The rigid body is said to be <u>non-degenerate</u> if the rank of the matrix (c_{ij}) defined by formula (5.1) is

equal to three, i.e., if the points $\vec{x}_1,\ldots,\vec{x}_N$ satisfying relations $(5,1)$ never lie in a lower dimensional affine submanifold of R^3.

Exercise. If the rigid body is non-degenerate, then G acts transitively on Q. Further, it acts simply, i.e., if an element g ε G leaves a point of G fixed, then g = identity.

The two properties provided by this exercise imply that:

> Q can be identified with G.

Namely, Q is the orbit of G. This identification enables us to describe the mechanics of rigid bodies in terms of Lie group-theoretic differential-geometric properties of G. This fact was first pointed out by V. Arnold [1] in a seminal paper, generalized by him to cover the theory of ideal fluids. For further ramifications, see Ebin and Marsden [1] and my paper "Geodesics and classical mechanics on Lie groups".

6.   THE HAMILTONIAN VERSION OF THE EQUATIONS OF MOTION

Return to Theorem 4.1, and the formula (4.10) which gives the coordinate-free version of Lagrange's equations.

We shall now show that Hamilton's equations result from choosing a <u>different coordinate system</u> for $T(Q) \times T = X$.

The variables with which we have been working up to now are the "Newtonian" coordinates

$$(q, \dot{q}, t) \quad .$$

The Lagrangian L is given as a function

$$L(q, \dot{q}, t)$$

of them.

Define new functions on X as follows:

$$p_i = \frac{\partial L}{\partial \dot{q}_i} \tag{6.1}$$

They are called the <u>momenta</u>.

<u>Definition</u>. The Lagrangian L is said to be <u>non-degenerate</u> if the functions

$$(q^i, p_j, t)$$

define a new coordinate system for X.

<u>Remark</u>. Here are alternate (and useful) versions of this condition:

$$dq^i, \ dp_i, \ dt \quad \text{are linearly independent.}$$

$$\det \frac{\partial^2 L}{\partial \dot{q}^i \partial \dot{q}^j} \neq 0$$

The map

$$T(Q) \times T \to T^d(Q) \times T$$

defined by the p_i (called the <u>Legendre transform</u>) is a local diffeomorphism.

Let us suppose the Lagrangian is non-degenerate. Let

$$\theta = L dt + \frac{\partial L}{\partial \dot{q}^i} (dq^i - \dot{q}^i dt) \qquad (6.2)$$

be the Cartan form associated with the Lagrangian L. With the functions p_i defined by (6.1), we can rewrite θ as follows:

$$\theta = p_i dq^i - H dt \qquad (6.3)$$

with:

$$H = p_i \dot{q}^i - L \qquad (6.4)$$

The function H is called the <u>Hamiltonian</u>.

As given by (6.4), H is a function of the variables $(q, \dot{q}, t)$. However, we can also express it as a function of the variables (q, p, t). Let us express it in this way. Let us also express the force differential form $\underline{F}$ in terms of these variables.

Now,

$$d\theta = dp_i \wedge dq^i - dH \wedge dt \quad . \qquad (6.5)$$

Let A be a vector field in the variables (q,p,t) such
that:

$$A(t) = 1 .$$

Then,

$$A \lrcorner d\theta = A(p_i)dq^i - A(q^i)dp_i - A(H)dt + dH$$

$$= \left(A(p_i) + \frac{\partial H}{\partial q^i}\right) dq^i + \left(\frac{\partial H}{\partial p_i} - A(q^i)\right) dp_i$$

$$+ \left(\frac{\partial H}{\partial t} - A(H)\right) dt .$$

Suppose the force form $\underline{F}$ is also expressed in terms of
these variables, in the following way:

$$\underline{F} = F_i(q,p,t)dq^i + F(q,p,t)dt . \qquad (6.6)$$

Then, we see that:

$$A \lrcorner d\theta = \underline{F}$$

if and only if the following conditions are satisfied:

$$A(p_i) = - \frac{\partial H}{\partial q^i} + F_i \qquad (6.7)$$

$$A(q^i) = \frac{\partial H}{\partial p_i} \qquad (6.8)$$

$$A(H) = \frac{\partial H}{\partial t} - F \qquad (6.9)$$

Now, the solutions of the Lagrange equations (4.10) are orbit curves of such a vector field A. We can then express the differential equations (4.10) in the following way:

Theorem 6.1. Let $t \to (q(t), p(t), t)$ be the coordinates of a curve in $T(Q) \times T$ which is a solution of the Lagrange equations (4.10). Then, these coordinates satisfy the following differential equation:

$$\frac{dq^i}{dt} = \frac{\partial H}{\partial p_i} (q(t), p(t), t)$$

$$\frac{dp_i}{dt} = -\frac{\partial H}{\partial q^i} (q(t), p(t), t) + F_i(q(t), p(t), t)$$

(6.10)

Remark. Equations (6.10) are called the _Hamiltonian version_ of the Lagrange equations.

Here is another important property of the Hamiltonian:

Theorem 6.2. If $H(q, p, t)$ is the Hamiltonian function associated with a non-degenerate Lagrangian $L(q, \dot{q}_i, t)$, then the following condition is satisfied:

$$\det \frac{\partial^2 H}{\partial p_i \partial p_j} \neq 0$$

(6.11)

Proof. We have:

$$H = p_i \dot{q}^i - L \tag{6.12}$$

In relation (6.12), give q and t <u>constant values</u>. With this understanding, apply the exterior differentiation operation d to both sides of (6.12):

$$\frac{\partial H}{\partial p_i} dp_i = \dot{q}^i dp_i + p_i d\dot{q}^i - p_i d\dot{q}^i$$

$$= \dot{q}^i dp_i \quad ,$$

or

$$\frac{\partial H}{\partial p_i} = \dot{q}^i \tag{6.13}$$

Now, originally the p_i were functions of the q^i (with q, t held constant, of course). "Non-degeneracy" of L means that this relation can be   nverted, and $\dot{q}^i$ expressed as a function of the p_i. The $\dot{q}^i$ are then to be regarded as functions of (p_i) whose <u>differentials are linearly independent</u>. Relation (6.13) then implies that relation (6.11) is satisfied.

Here is a converse to this result.

<u>Theorem 6.2</u>. Suppose that $H(q,p,t)$ is a function of the indicated variables such that relation (6.11) is satisfied.

Then there is a unique   non-degenerate Lagrangian function
$L(q,\dot{q},t)$ whose associated Hamiltonian function is H, i.e.,
is such that (6.11) is satisfied, with

$$p_i = \frac{\partial L}{\partial \dot{q}^i} \tag{6.14}$$

Exercise. Prove Theorem 6.2.

Remark. Here is a more global interpretation of this. Start
off with Q a manifold, with coordinates (q^i). L lives on
$T(Q) \times T$, with coordinates $(q^i, \dot{q}^i, t)$. Let $T^d(Q)$ denote
the cotangent bundle to Q. H lives on $T^d(Q) \times T$. Relation (6.14) can be interpreted as a map $T(Q) \times T \rightarrow T^d(Q) \times T$,
called the Legendre transform.

Example. Newtonian Lagrangian.

   Suppose that the Lagrangian L is of the following
form:

$$L = \frac{1}{2} g_{ij}(q)\dot{q}^i\dot{q}^j + h_i(q)\dot{q}^i - V(q) \tag{6.15}$$

Such a Lagrangian is said to be of Newtonian type.

   It is non-degenerate if and only if the following condition is satisfied:

$$\det (g_{ij}) \neq 0 \tag{6.16}$$

Now, we can suppose--without any loss in generality--that (g_{ij}) depends <u>symmetrically</u> on the indices (i,j). This, combined with (6.16), means that L has associated with it the following <u>Riemannian metric</u>:

$$ds^2 = g_{ij}dq^i dq^j . \qquad (6.17)$$

Let us calculate the Hamiltonian associated with the Lagrangian L.

$$p_i = \frac{\partial L}{\partial \dot{q}^i} = g_{ij}(q)\dot{q}^j \qquad (6.18)$$

Let

$$(g^{ij})$$

denote the inverse matrix to (g_{ij}), i.e.,

$$g^{ij}g_{jk} = \delta^i_k \qquad (6.19)$$

Thus, we can invert relation (6.18) as follows:

$$\dot{q}^i = g^{ij}p_j \qquad (6.20)$$

(Notice the virtues of the tensor analysis notation-- using superscripts as covariant indices, subscripts as contravariant indices!) Substitute (6.20) into (6.15):

$$L = \frac{1}{2} g^{ij}p_i p_j + h_i g^{ij}p_j - V \qquad (6.21)$$

Also

$$p_i \dot{q}^i = g^{ij} p_i p_j \qquad (6.22)$$

Put all this together to get:

$$H = p_i \dot{q}^i - L$$

$$= \frac{1}{2} g^{ij} p_i p_j - g^{ij} h_i p_j + V \qquad (6.23)$$

Formula (6.23) is the definitive formula. If $h_i = 0$ (which is a common situation), note that H is the _total_ _energy_ of the system. Then, the theorem that H is conserved corresponds physically to _conservation of energy_. This identification of the Hamiltonian with the total energy is also important in relativistic mechanics.

We can also write down the Lagrangian L and the Hamiltonian H in a coordinate-free way. Let Q be the configuration space manifold. Let

$$\beta: T(Q) \to R$$

be a mapping which defines a _Riemannian metric_ on Q, i.e., for each $q \in Q$, α is a non-degenerate quadratic form on Q_q. Let

$$\alpha: T(Q) \to R$$

be a map which is linear on the fibers, i.e., α is a one-differential form on Q. V is the pull-back to $T(Q)$ of a real-valued function on Q, called the _potential function_.

Regard L as a map

$$T(Q) \to R$$

Then, it is given as follows:

$$L = \frac{1}{2} \beta + \alpha - V \qquad\qquad (6.24)$$

Let $\beta^d: T^d(Q) \to R$ be the dual quadratic form. (What one might call a <u>co-Riemannian</u> or <u>contravariant Riemannian metric</u>.)

β also defines an isomorphism from $T(Q)$ to $T^d(Q)$. Let $\alpha^d: T^d(Q) \to R$ be the linear function resulting from carrying over α via this isomorphism. Then,

$$H = \frac{1}{2} \beta^d + \alpha^d + V \quad . \qquad\qquad (6.25)$$

These formulas might be interesting and useful in the study of mechanical systems with an infinite number of degrees of freedom, e.g., fields.

Chapter 8

ABELIAN GROUPS OF LAGRANGIAN SYMMETRIES AND
CYCLIC COORDINATES

1.   INTRODUCTION

A large part of the classical theory of analytical
mechanics is concerned with the appearance of what are called
"cyclic" or "ignorable" coordinates in a mechanics problem.
(See Whittaker [1], Pars [1] and Gantmacher [1].) Many of the
systems which are most commonly and usefully studied have this
property and they are typically the systems whose solutions
can be found by means of explicit formulas.

In this chapter, I will develop the group-theoretic
background to these phenomena. A "cyclic" coordinate is
usually associated with a one-parameter group of transformations
on the configuration space of the mechanical system which preserves the Lagrangian. Several cyclic coordinates correspond
to higher dimensional abelian transformation groups acting
on the configuration space.

I will first discuss the general aspects of this situation, then specialize to the sort of abelian groups that are
typical of the classical work.

2.   NOTATION AND DEFINITIONS

Let Q be a manifold. It is called the <u>configuration</u> <u>space</u>.

Suppose Q is an n-dimensional manifold. Choose indices and the summation convention as follows:

$$1 \le i,j \le n \quad,$$

Label a typical coordinate system for Q as:

$$(q^i) \quad.$$

We also use a vectorial notation

$$q \quad.$$

Thus, depending on the context, q can mean a point of Q or an n-vector, (q^i) representing the coordinates of a point of Q. Usually, the reader will have little difficulty understanding the meaning.

T(Q) denotes the tangent space to Q. Coordinates for Q may be labelled:

$$(q^i, \dot{q}^i) \quad.$$

They have the following meaning:

If q_0 is a point of Q, if $v \in Q_{q_0}$
is a tangent vector to Q at q_0, then:

$$q^i(v) = q^i(q_0)$$

$$\dot{q}^i(v) = dq^i(v) \quad.$$

(Recall that the differentials of functions, e.g., dq^i, are linear forms on tangent vectors.)

We use the following notation:

$(q,\dot{q})$ denotes a point of $T(Q)$.

Definition.   A Lagrangian for   Q   is a real valued function

$L\colon T(Q) \to R$.

We can then write   L   as

$L(q,\dot{q})$.

Definition.   A curve $t \to q(t)$ in   Q   is a solution of Lagrange's equations if:

$$\frac{d}{dt}\left(\frac{\partial L}{\partial \dot{q}^i}\, q(t),\, \frac{dq}{dt}\right) - \frac{\partial L}{\partial q^i}\left(q(t),\, \frac{dq}{dt}\right) = 0 \qquad (2.1)$$

Remark.   It would be possible to include "force" terms on the right hand side of (2.1). However, for simplicity, I will restrict attention in this chapter to the force-free case. t   usually denotes a real parameter varying over a fixed interval.

Let   G   be a group of diffeomorphisms of   Q.   G   admits a natural prolongation to be a transformation group on $T(Q)$:

$$g(v) = g_*(v) \qquad\qquad (2.2)$$

for $v \in T(Q)$,

where $g_* : T(Q) \to T(Q)$ is the natural map g_* induces on tangent vectors.

Remark. Here is what I mean. $F(Q)$ denotes the C^∞, real-valued functions on Q. A $v \in Q_q$ is an R-linear map: $F(Q) \to R$ such that

$$v(f_1 f_2) = v(f_1)f_2(q) + f_1(q)v(f_3)$$

for $f_1, f_2 \in F(Q)$.

Then, $g_*(v)$ is the element of $Q_{g(q)}$ defined as follows:

$$g_*(v)(f) = v(g^*(f)) \quad .$$

Here is another characteristic geometric property of this prolongation process. If

$$t \to q(t)$$

is a curve in Q, if

$$t \to q'(t) \in T(Q)$$

is its tangent vector curve, then

$$t \to g(q'(t))$$

is the tangent vector curve of the transformed curve

$$t \to g(q(t)) \quad .$$

<u>Definition</u>. G is said to <u>leave invariant the Lagrangian</u> L
<u>if</u>:

$$L(g(q,\dot{q})) \;=\; L(q,\dot{q}) \qquad\qquad (2.3)$$

for $g \; \varepsilon \; G$, $(q,\dot{q}) \; \varepsilon \; T(Q)$.

It is readily seen that condition (2.3) implies that G
<u>leaves invariant the family of solutions of the Lagrange</u>
<u>equations</u> (2.1), in the following sense:

If $t \rightarrow q(t)$ is a solution of (2.1),
then so is the transformed curve

$$t \rightarrow g(q(t))$$

for each $g \; \varepsilon \; G$.

One also says that G is a <u>group of Lagrangian symmetries</u>.

3.   CYCLIC COORDINATES AND ABELIAN GROUPS

Keep the notation of Section 2. The coordinates

$$(q^1,\ldots,q^m)$$

for Q are said to be <u>cyclic</u> or <u>ignorable</u> relative to the
Lagrangian L if:

$$\frac{\partial L}{\partial q^i} = 0 \tag{3.1}$$

for $1 \leq i \leq m$

In this case, the Lagrange equations (2.1) imply the following
conditions:

$$\frac{d}{dt}\left(\frac{\partial L}{\partial \dot{q}^i}\left(q(t), \frac{dq}{dt}\right)\right) = 0 \tag{3.2}$$

for $1 \leq i \leq m$, each solution curve $t \to q(t)$.

In words, the functions

$$\frac{\partial L}{\partial \dot{q}^i}$$

(which are <u>generalized momenta</u>) are <u>conserved</u> along the solutions of (2.1).

Group-theoretically, (3.1) can be interprelated as
follows. Let R^m be the additive group of real m-vectors
acting on Q as follows:

$$xq = (q^1 + x^1, \ldots, q^m + x^m, q^{m+1}, \ldots, q^n) \tag{3.3}$$

for

$$x = (x^1, \ldots, x^n) \; \varepsilon \; R^n$$

$$q = (q^1,\ldots,q^m, q^{m+1},\ldots,q^n) \in Q .$$

It is readily seen that condition (3.1) is equivalent to the following condition:

$$\boxed{\begin{array}{l} R^m \text{ acts as a group of Lagrangian} \\ \text{symmetries} \end{array}} \quad (3.4)$$

Condition (3.3) must often be interpreted in only a "local" way. Often, one is given a connected abelian, m-dimensional Lie group G acting on Q. (Typically, G will be an m-<u>torus</u>, i.e., the product of m copies of the planar rotation group $SO(2,R)$.) $q^1,\ldots,q^m$ are coordinates on this group. The physicist might say that "$q^1,\ldots,q^m$ are angular coordinates."

4.   EXAMPLES OF CYCLIC COORDINATES

The basic references for these examples are the treatises by Whittaker [1] and Pars [1].

Example 1.   <u>Central forces.</u>

$Q = R^2$ minus the origin. Let (x,y) denote Cartesian coordinates for R^2, and (r,θ) polar coordinates:

$$x = r \cos \theta$$

$$y = r \sin \theta \ .$$

$$L = \frac{1}{2} (\dot{r}^2 + r^2 \dot{\theta}^2) - V(r) \tag{4.1}$$

The angular coordinate θ is cyclic. The corresponding generalized momentum is:

$$\frac{\partial L}{\partial \dot{\theta}} = mr^2 \cdot$$

It is the <u>angular momentum</u>.

Choose coordinate

$$q^1$$

on Q so that

$$\tan q^1 = \frac{y}{x} \ .$$

Of course, q^1 is usually denoted by "θ", and we shall also use this notation.

$$L = \frac{1}{2} ma^2 \dot{\theta}^2 + mga \cos \theta \tag{4.2}$$

The group G of Lagrangian symmetries is, of course, SO(2,R), the group of <u>rotations</u> of R^2. m is the <u>mass</u> of the particle.

Example 2. The spherical pendulum.

A particle of mass m moves under the action of
gravity on a smooth sphere of radius a. θ is the sphere of
radius a in R^3. Choose the usual spherical coordinates as:

$$\theta, \phi$$

for Q. Then,

$$L = \frac{1}{2} ma^2(\theta^r + (\sin^2 \theta)\dot{\phi}^2) - mga \cos \theta \qquad (4.3)$$

ϕ is cyclic .

$$\frac{\partial L}{\partial \dot{\phi}} = ma^2(\sin^2\theta)\dot{\phi} \quad ,$$

which is a conserved quantity. m is the mass of the particle,
g is the gravitational constant.

Example 3. Spinning top (Pars, p. 113).

Q = SO(3,R) = group of 3×3 orthogonal matrices.
The coordinates are the three Euler angles

$$\theta, \quad \phi \quad \text{and} \quad \psi \quad .$$

$$L = \frac{1}{2} A(\dot{\theta}^2 + \dot{\phi}^2 \sin^2 \theta) + \frac{1}{2} C(\dot{\psi} + \dot{\phi} \cos \theta)^2 - Mg\ell \cos \theta$$

$$(4.4)$$

A and C are moments of inertia. M is the total mass of

the top, g is the gravitational constant. ℓ is the distan
from the center of mass of the top to the point about which
it rotates.

ϕ and ψ are cyclic coordinates. Here is their
group-theoretic origin.

Consider the subgroup

$$H = SO(3,R) \times SO(2,R)$$

$$\text{of } SO(3,R) \times SO(3,R) .$$

Let it act on Q via left and right translation, i.e.,

$$(g_1, g_2)(q) = g_1 q g_2^{-1}$$

for $g_1 \varepsilon SO(3,R)$, $g_2 \varepsilon SO(2,R)$, $q \varepsilon Q \equiv SO(3,$

("SO(2,R)" denotes the subgroup of rotations of the top <u>about</u>
<u>its axis of symmetry</u>.)   H, acting on Q, is the symmetry
group of the Lagrangian L.

$$G = SO(2,R) \times SO(2,R)$$

is a <u>maximal abelian subgroup</u> of H. It is this group which
gives rise to the cyclic coordinates

$$\phi, \psi$$

5.   THE ORBIT SPACE

         Return to the general situation. Q is a manifold,
G is a transformation group on Q. Assume that G is a
connected Lie group.

Remark. For the Lagrangians commonly occurring in physics,
one can prove that the group of symmetries is a Lie group.
Hence, this is a reasonable assumption.

         Let

              G\Q

be the orbit space of the action of G on Q. A point of
G\Q is then a subset of Q which is an orbit of G. In this
section, we make the following simplifying "non-singularity"
assumptions:

         a) The orbits of G acting on Q all have the
            same dimension.
         b) G\Q is a manifold.
         c) The quotient map

              $\pi: Q \to G\backslash Q$

            is a submersion mapping, i.e.,

              $\pi_*(T(Q)) = T(G\backslash Q)$.

(In fact, a typical group action will have "singularities".
They must be handled separately, in ways that will be indicat
later on.)

Denote a curve in Q by

$$\underline{q} \quad .$$

(Thus, $\underline{q}$ denotes a map $t \to q(t)$ of an interval of real
numbers to Q.)

The curve $\pi\underline{q}: t \to \pi(q(t))$ is called the <u>projection</u>
in G\Q.

Let $\underline{S}$ be the collection of the curves in Q which
are solutions of the Lagrange equations (2.1). Here is our
problem:

How does one describe the curves in G\Q
which are of the form $\pi(\underline{S})$, i.e., which
are projections under π of the solutions
of the Lagrange equations?

A sufficiently detailed answer to this question is only known
in case G is an <u>abelian</u> group. We shall consider this in
the next section.

6.   THE ROUTHIAN

Suppose now that G is an <u>abelian</u> Lie group acting
configuration space Q as a group of symmetries of the

Lagrangian L. We can then choose local coordinates for Q, which we label

$$(\phi^1,\ldots,\phi^m; \; \theta^1,\ldots,\theta^{n-m})$$

such that:

> L is independent of
> $\phi^1,\ldots,\phi^n$.

(6.1)

The $\phi^1,\ldots,\phi^m$ are said to be _ignorable_ or _cyclic coordinates_.

Choose indices and the summation convention as follows:

$$1 \leq a,b \leq m$$

$$1 \leq u,v \leq n-m$$

The vector fields

$$\frac{\partial}{\partial\phi^1},\ldots,\frac{\partial}{\partial\phi^m}$$

on Q then form the _Lie algebra of_ G. In particular, the functions

$$\theta^1,\ldots,\theta^{n-m}$$

are invariant under G, hence can be regarded as the pull-back under π of functions on

$$G\backslash Q \; .$$

We also denote these functions on $G\backslash Q$ by θ^u, hoping that the reader will be able to live with the ambiguity. Thus:

$$\theta^1, \ldots, \theta^{n-m}$$

are local coordinates for the orbit space $G\backslash Q$.

Introduce a new set of variables:

$$(p_1, \ldots, p_m) = p \quad.$$

Definition. A function

$$(\theta, \dot{\theta}, p) \rightarrow R(\theta, \dot{\theta}, p)$$

of the indicated variables is called a <u>Routhian</u> for L and the abelian symmetry group G if the following condition is satisfied:

$$\boxed{R\left(\theta, \dot{\theta}, \frac{\partial L}{\partial \dot{\phi}}\right) = L - \dot{\phi}^a \frac{\partial L}{\partial \dot{\phi}^a}} \qquad (6.1)$$

We shall suppose that the Lagrangian L satisfies the following non-degeneracy condition:

The function

$$\phi^a, \frac{\partial L}{\partial \dot{\phi}^a}, \theta^u, \dot{\theta}^u \qquad (6.2)$$

are functionally independent, i.e.,
their differentials are everywhere
linearly independent.

Theorem 6.1 (Routh). Let $t \to q(t) \equiv (\phi(t), \theta(t))$ be a solution of the Lagrange equations, (2.1). Let:

$$p_a = \left(\frac{\partial L}{\partial \dot{\phi}^a} \left(q(t), \frac{dq}{dt} \right) \right) \tag{6.3}$$

(Recall that they are constants, since the coordinates ϕ are cyclic.) Then, the projected curve $t \to \pi \dot{q}(t)$ in $G \backslash Q$ has coordinates

$$\theta^u(t)$$

It satisfies the following differential equations:

$$\frac{d}{dt} \left(\frac{\partial R}{\partial \dot{\theta}^u} \left(\theta(t), \frac{d\theta}{dt}, p \right) \right) = \frac{\partial R}{\partial \theta^u} \left(\theta(t), \frac{d\theta}{dt}, p \right) \tag{6.4}$$

Proof. Cosider the space of variables

$$(\phi, \dot{\phi}, \theta, \dot{\theta}, t) \quad .$$

Introduce new coordinates for this space as follows:

$$\left(\phi, p_a \equiv \frac{\partial L}{\partial \dot{\phi}^a}, \theta, \dot{\theta}, t \right)$$

Now,

$$dL = \frac{\partial L}{\partial \dot{\phi}^a} d\dot{\phi}^a + \frac{\partial L}{\partial \theta^u} d\theta^u + \frac{\partial L}{\partial \dot{\theta}^u} d\dot{\theta}^u$$

$$= p_a d\dot{\phi}^a + \frac{\partial L}{\partial \theta^u} d\theta^u + \frac{\partial L}{\partial \dot{\theta}^u} d\dot{\theta}^u$$

$$d \left(\phi^a \, \frac{\partial L}{\partial \dot{\phi}^a} \right) \;=\; p_a d\phi^a \,+\, \phi^a dp_a$$

Hence,

$$dR \;=\; \frac{\partial L}{\partial \theta^u} \, d\theta^u \,+\, \frac{\partial L}{\partial \dot{\theta}^u} \, d\dot{\theta}^u \,-\, \phi^a \, dp_a \tag{6.5}$$

This relation tells us, first, that R is a function, in
these new coordinates, of $(\theta,\dot{\theta},p)$ <u>alone</u>. Second, the partial
derivatives of R <u>with respect to these new variables</u> are
given as follows:

$$\frac{\partial R}{\partial \theta^u} \;=\; \frac{\partial L}{\partial \theta^u}$$

$$\tag{6.6}$$

$$\frac{\partial R}{\partial \dot{\theta}^u} \;=\; \frac{\partial L}{\partial \dot{\theta}^u}$$

Substituting these relations into the Lagrange equations (2.1)
gives the equations (6.4).

<u>Remark</u>. Keep in mind the conventions inherent in (6.6)--
otherwise the notation is very confusing. On the left hand
side, the partial derivatives are calculated as follows:
First, express R as a function of $\theta,\dot{\theta},p$, then calculate
derivatives. For the right hand side, calculate derivatives
of L as a function of the old coordinates $\phi,\theta.\dot{\theta}$.

Exercise. Prove the converse of Theorem 6.1, i.e., that if
$p = (p^a)$ is an element of R^m, and

$$t \rightarrow \theta(t)$$

a curve in R^{n-m} which solves (6.4), there is a curve

$$t \rightarrow (\phi(t), \theta(t))$$

in $R^m \times R^{n-m}$ which solves the Lagrange equations (2.1),
and such that:

$$p_a = \frac{\partial L}{\partial \dot{\phi}^a} \left(\frac{d\phi}{dt}, \theta(t), \frac{d\theta}{dt} \right) .$$

Let us recapitulate what has been done in a more
global language. Recall that Q is a manifold, on which an
abelian, connected Lie group G acts. This action is non-
singular, in the sense that the orbit space

$$G \backslash Q$$

is a manifold, and the quotient map

$$\pi: Q \rightarrow G \backslash Q$$

is a submersion.

(ϕ, θ) are coordinates on Q. (θ) are the pull-back
under π of coordinates on the orbit space.

$T(Q)$ denotes the tangent vector bundle to Q. The
"submersion" property for π means that

$$\pi_*: T(Q) \to T(G\backslash Q)$$

is <u>onto</u>.

L is a real-valued function on $T(Q)$. Suppose that it is a non-degenerate Lagrangian. There is then a vector field A on $T(Q)$ whose orbit curves are the solutions of the Lagrange equations (2.1).

The momenta $p_a = \partial L/\partial \dot{\phi}^a$ are functionally independent functions on $T(Q)$, and, when set equal to constants, determine a foliation of $T(Q)$. The orbit curves of A, i.e., the solution of the Lagrange equations, lie on the leaves of this foliation. For each leaf of the foliation, the Routhian defines a function on $T(G\backslash Q)$, i.e., a Lagrangian for the orbit space $G\backslash Q$. The orbit curves of A, which lie on a given leaf, then define, when projected via π to $G\backslash Q$, the family of solutions of the Lagrange equations corresponding to a Routhian.

<u>Remark</u>. This interpretation fits in quite well with my general comments on Lie's work on symmetries of differential equations in the "Lie group" series, Vol. 3. There, I pointed out that much of Lie's work could be thought of as investigating the "structure" of the "orbit space" of a group of symmetries acting on the <u>space of solutions of a differential equation</u>. Notice that Routh's idea is a variant of Lie's.

7.   THE ROUTHIAN FOR LAGRANGIANS OF NEWTONIAN TYPE

Let us now specialize the situation considered in Section 6 so that it has the following form:

$$L = \frac{1}{2} g_{ab}(\theta) \dot{\phi}^a \dot{\phi}^b + \frac{1}{2} g_{uv}(\theta) \dot{\theta}^u \dot{\theta}^v + g_{au}(\theta) \dot{\phi}^a \dot{\theta}^u + g_a(\theta) \dot{\phi}^a \qquad (7.1)$$

$$+ g_u(\theta) \dot{\theta}^u + V(\theta)$$

This is a Lagrangian of <u>Newtonian type</u>. (Typically, problems of Newtonian mechanics lead to Lagrangians of this type.) As the notation in (7.1) indicates, the coefficients in (7.1) are functions of θ alone, i.e., the coordinates ϕ are <u>cyclic</u>. We shall now calculate the Routhian explicitly.

$$p_a = \frac{\partial L}{\partial \dot{\phi}^a} = g_{ab} \dot{\phi}^b + g_{au} \dot{\theta}^u \qquad (7.2)$$

Let (g^{ab}) denote the converse matrix to (g_{ab}). Then,

$$\dot{\phi}^a = g^{ab} p_b + g^{ab} g_{bu} \dot{\theta}^u \qquad (7.3)$$

$$p_a \dot{\phi}^a = g_{ab} \dot{\phi}^a \dot{\phi}^b + g_{au} \dot{\theta}^u \dot{\phi}^a$$

$$R = L - p_a \dot{\phi}^a = -\frac{1}{2} g_{ab} \dot{\phi}^a \dot{\phi}^b + \frac{1}{2} g_{uv} \dot{\theta}^u \dot{\theta}^v + g_a \dot{\phi}^a + g_u \dot{\theta}^u + V(\theta)$$

$$g_{ab} \dot{\phi}^a \dot{\phi}^b = \dot{\phi}^a (p_a - g_{au} \dot{\theta}^u) = (g^{ab} p_b - g^{ab} g_{bu} \dot{\theta}^u)(p_a - g_{au} \dot{\theta}^u)$$

$$= g^{ab} p_a p_b - 2 g^{ab} g_{bu} \dot{\theta}^u p_a + g^{ab} g_{bu} g_{av} \dot{\theta}^u \dot{\theta}^v$$

Hence,

$$R = \frac{1}{2} g_{uv} \dot{\theta}^u \dot{\theta}^v - \frac{1}{2} g^{ab} g_{bu} g_{av} \dot{\theta}^u \dot{\theta}^v - \frac{1}{2} g^{ab} p_a p_b + g^{ab} g_{bu} \dot{\theta}^u p_a$$

$$(7.4)$$

$$+ g_a (g^{ab} p_b + g^{ab} g_{bu} \dot{\theta}^u) + g_u \dot{\theta}^u + V(\theta)$$

8.   CONDITIONS FOR AN EQUILIBRIUM POINT ON THE QUOTIENT

Return to the general case of a Lagrangian

$$L: T(Q) \to R \quad,$$

with coordinates (ϕ, θ) such that the ϕ are cyclic, i.e., L is a function of $(\dot{\phi}, \theta, \dot{\theta})$ alone. The coordinates ϕ are cyclic. Let

$$p = \frac{\partial L}{\partial \dot{\phi}}$$

be the corresponding conserved momenta.

Definition. A pair (p, θ_0), $p \in R^m$, $\theta \in G\backslash Q$, is said to define a quotient equilibrium point of the system if, there i. a solution $t \to (\phi(t), \theta(t))$ of the Lagrange equations such that:

$$\theta(t) = \theta_0 \qquad\qquad\qquad (8.1)$$

$$p = \frac{\partial L}{\partial \dot{\phi}} \left(\frac{d\phi}{dt}, \theta(t), \frac{d\theta}{dt} \right) \qquad (8.2)$$

In words, the quotient coordinates θ remain constant,
while the conserved momenta have the values p. In the classical analytical mechanics literature, such a solution is often
called a <u>steady motion</u> of the system.

<u>Remark</u>. The study of such equilibria is a main theme of the
classical work on cyclic coordinates. One is particularly
interested in the <u>stability</u> properties of such equilibria.
See the extensive work in Whittaker, Pars and Gantmacher.

Here is a main property.

<u>Theorem 8.1</u>. Suppose (p, θ_0) is a quotient equilibrium point
of the system. Suppose also that the Lagrangian satisfies
the following non-degeneracy condition:

$$\det \left(\frac{\partial^2 L}{\partial \dot{\phi}^a \partial \dot{\phi}^b} \right) \neq 0 \tag{8.3}$$

Then,

$$\frac{d^2}{dt^2} \phi(t) = 0 \tag{8.4}$$

<u>Proof</u>. Use the fact that

$$t \rightarrow (\phi(t), \theta_0)$$

is a solution of Lagrange's equations:

$$\frac{d}{dt} \left(\frac{\partial L}{\partial \dot{\theta}^u} \right) = \frac{\partial L}{\partial \theta^u}$$

$$\frac{d}{dt} \left(\frac{\partial L}{\partial \dot{\phi}^a} \right) = \frac{\partial L}{\partial \phi^a}$$

(8.5)

Recall that L is a function

$$L(\dot{\phi}, \theta, \dot{\theta})$$

of the indicated variables. Then, (8.4) takes the following form: (using the fact that $\theta(t) \equiv$ constant),

$$\frac{\partial^2 L}{\partial \dot{\phi}^a \partial \dot{\phi}^b} (\phi, \theta_0, 0) \frac{d^2 \phi^b}{dt^2} = 0 \qquad (8.6)$$

The non-degeneracy condition (8.3) together with (8.6) implies (8.4).

Remark. Here is a physical meaning of condition (8.4). It implies that $\phi^a(t)$ has the following form:

$$\phi^a(t) = \alpha^a t + \beta$$

Thus, the functions

$$e^{\phi^a}$$

will exhibit typical oscillatory and/or decaying behavior. This "equilibrium" is then of a sort that certain coordinates (the cyclic ones) have this oscillatory behavior, while the

other coordinates--namely those which are invariant under the
abelian symmetry group G --remain constant. A typical
example is a symmetric heavy top. θ is the Euler angle
about the (vertical) axis of symmetry. ϕ^1, ϕ^2 are the other
Euler angles, which exhibit this typical oscillatory behavior.

I mean to come back to these stability problems--perhaps
even in a more general setting--in a later volume.

Chapter 9

SOME RELATIONS BETWEEN ANALYTICAL
AND QUANTUM MECHANICS

1.   INTRODUCTION

As I have already emphasized in several of my books, understanding the many mathematical facets of quantum mechanics is
the greatest challenge to contemporary mathematical physics.
My particular concern has always been the relation between
quantum and "classical" (i.e., "analytical") mechanics. Physicists have tended, in recent years, to minimize this relation,
but I believe that this has been a mistake.

On the mathematical side, these relations have also recently
been emphasized by Kostant [1] and Souriau [1], following
earlier work by Van Hove [1] and Segal [1]. Unfortunately, this
work on the "geometry of quantization" has not considered deeply
the physical problem, although it certainly is of mathematical
interest (I have in mind the comments of Van Hove [1] and
Prosser [1] about this method of quantization--papers which one
never finds referred to in this recent work!)

In my previous work (e.g., VB, Vol. 2, LAQM, PALG and IM,
Vol. 6) I have attempted to begin a systematization of the
geometric and Lie-group theoretic side of the quantization
problem. Although I am not yet prepared to write the systematic

treatise that is so obviously needed, I continue in this
chapter the preparatory work for this task, emphasizing
the relation to "analytical mechanics".

Customarily, the theory of Hilbert spaces and associated
ideas of functional analysis are considered as the fundamenta
mathematical background to quantum mechanics. Many contemporary papers on mathematical physics appear to be exercises
in exceedingly abstract functional analysis! In fact, a good
deal of the most interesting mathematical aspects of quantum
mechanics involves the geometric theory of differential operators on manifolds, and is strongly related to analytical
mechanics. I have already (e.g., in VB, Vol. 2) discussed
the interrelation between quantum mechanics, the theory of
differential operators and analytical mechanics. In this
chapter I continue this development.

2. DIFFERENTIAL OPERATORS, SYMBOLS AND POISSON BRACKET

Let Q be a manifold. (As the notation "Q" should indicate, Q will be identified with the "configuration space" of a mechanical system.)

As in previous chapters, local coordinate systems for Q will be denoted by

$$(q^i) ,   1 \leq i,j \leq n = \dim Q .$$

(Recall that n is called the <u>number of degrees of freedom of the mechanical system</u>.)

> $F_0(Q,C)$ denotes the C^∞, complex-valued functions on Q of <u>compact support</u>. (A function has compact support if it vanishes outside some compact subset of Q.)

<u>Definition</u>. A mapping

$$\Delta: F_0(Q,C) \to F_0(Q,C)$$

is a <u>linear differential operator</u> of <u>order</u> r if, in each coordinate system (q^i), Δ has the following form:

$$\Delta = a^{i_1 \cdots i_r} \frac{\partial^r}{\partial q^{i_1} \cdots \partial q^{i_r}} + \cdots \tag{2.1}$$

where:

The $(a^{(\ )}, \dots)$ are complex-valued, C^{∞}
functions of the coordinates fixed by Δ
and the choice of coordinates. The terms
... indicate similar differentiations of
order less than r.

Remark. There are more abstract ways to define the "linear
differential operator" concept. From the point of view of
functional analysis, the most basic is that by Peetre: A C^{∞}-
linear map on functions, which is "local", in the sense that,
if f vanishes in an open subset of Q so does $\Delta(f)$. For
differential-geometric purposes, the algebraic definition
given in GPS, Chapter 1, is more useful.

Let $T(Q)$ denote the tangent vector bundle to Q. Let
$T^d(Q)$ denote the cotangent bundle, i.e., the fiber to each
point $q \in Q$ is the dual space Q_q^d to the tangent space to Q.
A one-differential form on Q is then identified with a cross-
section of $T^d(Q)$.

A coordinate system (q^i) for Q defines a coordinate
system for $T^d(Q)$, that we label as

$$(q^i, p_j) \quad .$$

This will be called a canonical coordinate system for $T^d(Q)$. He
is how it is defined:

For $q_0 \, \varepsilon \, Q$, $\theta \, \varepsilon \, Q_{q_0}^d$,

$$q^i(\theta) \;=\; q^i(q_0)$$

$$p_i(\theta) \;=\; \theta(\partial/\partial q_0^i) \quad .$$

In words, q^i are the coordinates on θ, lifted up to $T^d(Q)$ via the pull-back of the projection map, and p_i are the duals to the "differentials" dq^i.

$T^d(Q)$ has an invariantly defined <u>contact stucture</u>--namely, that defined by the one-form $p_i dq^i$--and a <u>symplectic structure</u>, that defined by

$$\omega \;=\; dp_i \wedge dq^i \qquad\qquad (2.2)$$

Let $F(Q,C)$ denote the complex-valued, C^∞ functions on Q. Let $F(Q)$ denote the subset consisting of the <u>real valued</u> functions. The symplectic form ω given by (2.2) defined a Lie algebra structure on $F(Q,C)$, called <u>Poisson bracket</u>, defined as:

$$\{f_1,f_2\} \;=\; \frac{\partial f_1}{\partial p_i}\frac{\partial f_2}{\partial q^i} \;-\; \frac{\partial f_2}{\partial p_i}\frac{\partial f_1}{\partial q^i}$$

<u>Definition</u>. Let $\Delta: F_0(Q,C) \to F_0(Q,C)$ be a linear differential operator of form (2.1). The <u>symbol of</u> Δ is a complex-valued function

$$\sigma_\Delta : \; T^d(Q) \to C$$

defined, in canonical local coordinates, by the following
formula:

$$\sigma_\Delta(q,p) \;=\; a^{i_1 \cdots i_r}(q) p_{i_1} \cdots p_{i_r} \tag{2.}$$

Remark. This definition can be given in a coordinate-free wa
See VB and GPS. Notice particularly that the right hand side
of (2.2) does not involve the terms in Δ indicated by ...
in (2.1), i.e., which are of order lower than r. The point
is that these terms do not transform "tensorially".

Theorem 2.1. Let Δ_1, Δ_2 be two differential operators
$: F_0(Q,C) \to F_0(Q,C)$. Let $[\Delta_1, \Delta_2]$ be their commutator, i.e.,

$$[\Delta_1, \Delta_2] \;=\; \Delta_1 \Delta_2 - \Delta_2 \Delta_1 \;.$$

Then,

$$\sigma_{[\Delta_1, \Delta_2]} \;=\; \{\sigma_{\Delta_1}, \sigma_{\Delta_2}\} \tag{2.}$$

$$\sigma_{\Delta_1 \Delta_2} \;=\; \sigma_{\Delta_1} \sigma_{\Delta_2} \tag{2.}$$

For the proof of (2.3) and (2.4) (which may, in fact, be
done by a straightforward computation), see VB, Vol. II.

Warning. There is a notational convention inherent in (2.3)
and (2.4) which may be confusing. The symbol σ_Δ really depends
on Δ and the integer r. Thus, Δ may be an operator of
degree (r-1); then $\sigma_{\Delta,r} = 0$, but $\sigma_{\Delta,r-1}$ may not be zero.
Thus, (2.3) should be written:

$$\sigma_{[\Delta_1,\Delta_2],r+s-1} = \{\sigma_{(\Delta_1,r)},\sigma_{(\Delta_2,s)}\} \quad ,$$

where Δ_1 is order r, Δ_2 is order s. Similarly, (2.4) is:

$$\sigma_{\Delta_1\Delta_2,rs} = \sigma_{\Delta_1,r}\sigma_{\Delta_2,s}$$

However, I will not carry along this precise notation, leaving
it to the reader to keep it in mind.

Examples

a)   r = s = 0

Δ_1,Δ_2 are multiplication by functions

$$a_1(q), a_2(q) \quad .$$

Then,

$$\sigma_{\Delta_1} = a_1, \quad \sigma_{\Delta_2} = a_2$$

$$[\Delta_1,\Delta_2] = 0 = \{\sigma_{\Delta_1},\sigma_{\Delta_2}\}$$

b) $r = 1$, $s = 0$

$$\Delta_1 = a_1^i(q) \frac{\partial}{\partial q^i} \quad , \quad \Delta_2 = a_2(q)$$

$$[\Delta_1, \Delta_2] = a_1^i \frac{\partial a_2}{\partial q^i}$$

$$\sigma_{\Delta_1} = a_1^i p_i \quad , \quad \sigma_{\Delta_2} = a_2$$

$$\{\sigma_{\Delta_1}, \sigma_{\Delta_2}\} = a_1^i \frac{\partial a_2}{\partial q^i} = \sigma_{[\Delta_1, \Delta_2]} \quad .$$

c) $r = 1$, $s = 1$

$$\Delta_1 = a_1^i(q) \frac{\partial}{\partial q^i} \quad , \quad \Delta_2 = a_2^j(q) \frac{\partial}{\partial q^j}$$

Δ_1, Δ_2 are vector fields.

$$[\Delta_1, \Delta_2] = \left(a_1^i \frac{\partial a_2^j}{\partial q^i} - a_2^i \frac{\partial a_1^j}{\partial q^i} \right) \frac{\partial}{\partial q^j}$$

is the Jacobi bracket, again a vector field. Thus,

$$\sigma_{[\Delta_1, \Delta_2]} = \left(a_1^i \frac{\partial a_2^j}{\partial q^i} - a_2^i \frac{\partial a_1^j}{\partial q^i} \right) p_j$$

$$\{\sigma_{\Delta_1}, \sigma_{\Delta_2}\} = \{a_1^i p_i, \ a_2^j p_j\}$$

$$(2.5)$$

$$= a_1^i \frac{\partial a_2}{\partial q^i} p_j - a_2^j \frac{\partial a_1^i}{\partial q^j} p_i$$

They become equal when the substitution

$$j \to i$$
$$i \to j$$

is made on the right hand side of (2.5).

<u>Exercise.</u> If $r = 2$, $s = 0, 1$ or 2, work out explicitly the relation between operator commutator and Poisson bracket.

3.   THE ADJOINT OF A DIFFERENTIAL OPERATOR

Continue with the notation of Section 2. Q is a manifold, $F_0(Q,C)$ denotes the C^∞, compact support differential operator, etc. In addition, suppose that Q is an <u>orientable manifold</u>. (See DGCV.) Fix a volume element differential form on Q, denoted by

$$dq \quad .$$

(Recall that dq is a differential form of degree $n = \dim Q$ which is <u>non-zero</u> at each point of Q.)

Thus, if $f \in F_0(Q,C)$ the complex number

$$\int_Q f \, dq \equiv \int_Q f(q) \, dq \qquad (3.1$$

is defined and called the <u>integral of</u> f <u>over</u> Q.

<u>Remark</u>. Here is the reason I have chosen to deal with "compac support" functions--their integral always is finite. It is appropriate for various purposes to replace "compact support" with "rapidly decreasing". In one of my papers [12], I have done this for general manifolds, but the subject is somewhat too technical and complicated to go into at this point.

We can now use the volume element (3.1) to put an inner product, denoted by $< \mid >$, on $F_o(Q,C)$:

$$<f_1 | f_2> = \int_Q f_1(q)^* f_2(q) \, dq \qquad (3.2$$

(* denotes the <u>complex conjugate</u> operation.)

The inner product has the following properties:

It is R-linear

$$<cf_1 | f_2> = c^* <f_1 | f_2> = <f_1 | c^* f_2>$$

for $c \in C$.

$$<f_1 | f_2>^* = <f_2 | f_1>$$

$$<f | f> > 0$$

These algebraic properties may be summed up by saying that they define a Hermitian symmetric space. A complex vector space carrying a complex-valued inner product with these properties may be called a Hermitian space.

Remark. In my previous work (e.g., FA, VB, PALG) I have proposed calling a space with these properties a Hilbert space, since this is basically the terminology in physics. (The mathematicians add "completeness.") Since this seems to be misunderstood, I will replace "Hilbert" by "Hermite", reserving the name "Hilbert space" for a complete Hermitian space.

Now, each differential operator Δ defines a C-linear map $F_0(Q,C) \to F_0(Q,C)$.

Theorem 3.1. There is a map

$$\Delta \to \Delta^*$$

on differential operators such that:

$$\langle f_1 | \Delta f_2 \rangle = \langle \Delta^* f_1 | f_2 \rangle \qquad (3.3)$$

$$\text{for} \quad f_1, f_2 \in F_0(Q,C) \quad ,$$

where $\langle \,|\, \rangle$ is the Hermitian inner product defined by (3.2). This map is called the (Hermitian) adjoint map. It has the following algebraic properties:

It is R-linear

$(\Delta^*)^* = \Delta$

$(c\Delta)^* = c^* \Delta$

   for $c \in C$.

$(\Delta_1 \Delta_2)^* = \Delta_2^* \Delta_1^*$

If Δ_1 is differential operator such that

   $\langle \Delta f_1 | f_2 \rangle = \langle f_1 | \Delta_1 f_2 \rangle$

     for all $f_1, f_2 \in F_0(Q,C)$,

then $\Delta_1 = \Delta^*$, i.e., Δ^* is <u>characterized</u>
by Property (3.3).

(3.4

<u>Proof</u>. I shall <u>sketch</u> the proof only in case Q has a
<u>globally defined</u> coordinate system (q^i), $1 \le i \le n$. The
general case will then follow using the "patching together"
(i.e., "partition of unity") arguments which are familiar in
differential geometry.

   Suppose that:

$$dq = \rho dq^1 \wedge \cdots \wedge dq^n ,$$

where ρ is a function. Since dq is non-zero, ρ is non-ze
We can suppose (at most reordering the variables) that

$\rho > 0$.

Thus, the adjoint operator Δ^* is characterized by the following (proposed) identity involving ordinary integrals over R^n:

$$\int f_1^* \Delta(f_2)(q)\rho(q)dq^1 \wedge \ldots \wedge dq^n = \int (\Delta^*(f_1))^* f_2(q)\rho(q)dq^1 \wedge \ldots \wedge dq^n ,$$

(3.5)

which is to hold for each pair (f_1, f_2) of functions on R^n which vanish for $q \in R^n$ sufficiently large. Δ has the following form:

$$\Delta = a^{i_1 \ldots i_r} \frac{\partial^r}{\partial q^{i_1} \ldots \partial q^{i_N}} + \cdots$$

(3.6)

An explicit solution for Δ^* is now evident:

$$\Delta^*(f) = \frac{(-1)^r}{\rho} \frac{\partial^r}{\partial q^{i_1} \ldots \partial q^{i_N}} \left(a^{i_1 \ldots i_r} \rho f \right) + \cdots$$

(3.7)

The proof of properties (3.4) is routine, and is left to the reader.

Remark. Although it is not needed explicitly at the moment, it is useful to keep in mind that (3.3) is a special case of a formula involving manifolds with boundary. Let Q be a compact orientable manifold with oriented boundary ∂Q. I will assume that the reader knows what these terms mean. Here is

a picture which illustrates the situation in the two dimensional case:

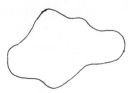

Let f_1, f_2 be C^∞, complex valued functions which are
C^∞ on Q and up to and including the boundary. Let $\vec{n}$ be
a vector field on Q which is transversal to the boundary, and
points inward.

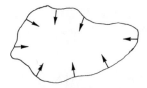

(There is no intrinsic "metric", so that one cannot say--as
in vector analysis--that $\vec{n}$ is the "unit normal", but it does
play the same role.)

Let dq be a volume element for Q. Then

$$\vec{n} \lrcorner dq$$

is a volume element form for the boundary manifold ∂Q. Suppose
that Δ is a linear differential operator on Q, which is C^∞
in terms of the boundary coordinates as well. Here is the
formula generalizing (3.3).

$$\int_Q (f_1^* \Delta(f_2) dq - (\Delta^* f_1^*) f_2 \, dq = \int_{\partial Q} \theta(f_1, f_2)(\vec{n} \lrcorner \, dq) \qquad (3.8)$$

Here θ is a bilinear differential operator

$$F(\partial Q, C) \otimes F(\partial Q, C) \to F(\partial Q, C) \quad .$$

This general formula readily specializes to many of the integration formulas used in mathematical physics and differential
equation theory. For example, the Lagrange adjoint formula of
linear ordinary differential equation theory, the Green's
formula of potential theory, Kirchoff's formulas of electro-
magnetic theory, etc.

Here are the definitions of concepts which are important
for quantum mechanics.

Definition. Let Δ be a differential operator on Q. It is
said to be:

> Hermitian if $\Delta^* = \Delta$
>
> Skew-Hermitian if $\Delta^* = -\Delta$

Note the following facts:

> Δ is Hermitian if and only if $i\Delta$ is skew-
> Hermitian. Δ can be split up into a sum of
> Hermitian and anti-Hermitian parts:

$$\Delta = \frac{1}{2} (\Delta + \Delta^*) + \frac{1}{2} (\Delta - \Delta^*)$$

If Δ_1, Δ_2 are skew-Hermitian, then:

$$[\Delta_1, \Delta_2] \equiv \Delta_1 \Delta_2 - \Delta_2 \Delta_1 \tag{3.9}$$

is skew-Hermitian

$$[\Delta_1, \Delta_2]_+ \equiv \frac{1}{2} (\Delta_1 \Delta_2 + \Delta_2 \Delta_1) \tag{3.1}$$

is Hermitian

Thus, the set of skew-Hermitian operators (which play the role of "quantum observables" and "infinitesimal generators of quantum symmetries") have two sorts of algebraic structures: A Lie algebra (3.9) and a Jordan algebra (3.10).

Examples.

First, suppose

$$\Delta = a^i \frac{\partial}{\partial x^i} \quad ,$$

i.e., Δ is a first order differential operator, i.e., a vector field. Then,

$$\Delta^* = - \frac{\partial}{\partial x^i} a^{i*}$$

$$= -a^{i*} \frac{\partial}{\partial x^i} - \frac{\partial (a_i^*)}{\partial x^i} \quad .$$

For example, if the a^i are <u>real</u>, we have:

$$\Delta^* = -\Delta - \text{div} (\Delta)$$

In particular,

$$\Delta - \frac{1}{2} \text{div} (\Delta)$$

is <u>skew-Hermitian</u>.

Suppose now that Δ is <u>real</u> and of second degree:

$$\Delta = a^{ij} \frac{\partial^2}{\partial x^i \partial x^j}$$

Then,

$$\Delta^* = \frac{\partial^2}{\partial x^i \partial x^j} a^{ij}$$

$$= \Delta + 2 \frac{\partial a^{ij}}{\partial x^i} \frac{\partial}{\partial x^j} \quad \frac{\partial^2 (a^{ij})}{\partial x^i \partial x^j}$$

Let us work out the conditions that real operators of the form

$$\Delta = a^{ij} \frac{\partial^2}{\partial x^i \partial x^j} + b^i \frac{\partial}{\partial x^i}$$

be Hermitian:

$$\Delta^* = a^{ij} \frac{\partial^2}{\partial x^i \partial x^j} + 2 \frac{\partial a^{ij}}{\partial x^i} \frac{\partial}{\partial x^j} + \frac{\partial^2 (a^{ij})}{\partial x^i \partial x^j} - b^i \frac{\partial}{\partial x^i} - \frac{\partial b^i}{\partial x^i} \quad .$$

In order that Δ be Hermitian, this must equal Δ. Compare
coefficients:

$$b^i = 2 \frac{\partial a^{ji}}{\partial x^j} - b^i \tag{3.1}$$

$$\frac{\partial^2 (a^{ij})}{\partial x^i \partial x^j} = \frac{\partial b^i}{\partial x^i} \tag{3.1}$$

Notice that (3.12) follows from (3.11).

We conclude that given a "pure" second order differential
operator,

$$\Delta = a^{ij} \frac{\partial^2}{\partial x^i \partial x^j} \quad ,$$

there is just one way to add a first order operator $b^i (\partial/\partial x^i$
to it to make it Hermitian. This observation plays a key role
in quantization of the "energy" in quantum mechanics.

Notice also that the Hermitian operator can be written a

$$\frac{\partial}{\partial q^i} a^{ij} \frac{\partial}{\partial q^j} \quad .$$

Now, let us return to the relation between quantum and
classical mechanics.

4. A FORMULATION OF THE MEANING OF "QUANTIZATION" OF A MECHANICAL SYSTEM

Let Q be a manifold, the <u>configuration space</u> of a mechanical system. The cotangent vector bundle

$$T^d(Q) \ = \ M$$

is called the <u>phase space</u> of the system. (Apparently, this definition is due to Gibbs. It is also the "state space" in the sense of systems theory.) The usual symplectic form on $M \equiv T^d(Q)$ defines a Poisson bracket structure { , } on $F(M)$, the space of real-valued, C^∞ functions on M.

<u>Definition</u>. A system of <u>classical observables</u> for the system, denoted by $\underline{0}$, is defined as a set of real-valued, C^∞ functions on M satisfying the following conditions:

a) $\underline{0}$ is a real vector space

b) $\underline{0}$ is closed under Poisson bracket

c) $\underline{0}$ <u>separates points</u> on M, i.e., if p_1, p_2 are two points on M such that

$$f(p_1) \ = \ f(p_2) \qquad \text{for all } f \ \varepsilon \ \underline{0},$$

then $p_1 = p_2$.

<u>Definition</u>. Let $\underline{0}$ define a system of classical observables. Let

$$\underline{\text{LD}}(Q)$$

be the space of linear differential operators $\Delta: F_0(Q,C) \to F_0(Q,C)$

Then, a _quantization_ of $\underline{0}$ is defined as an R-linear map

$$\underline{0} \to \underline{LD}(Q) \quad ,$$

denoted by

$$f \to \Delta_f \quad ,$$

such that the following conditions are satisfied:

The symbol of (Δ_f) is cf_1 where c is a complex scalar.

Δ_f is skew-Hermitian

$$[\Delta_{f_1}, \Delta_{f_2}] = \Delta_{\{f_1,f_2\}}$$

i.e., the map is a Lie algebra homomorphism.

Remark. The physicists describe this in a different way. They require that

$$f \to \Delta_f$$

be a map into _Hermitian_ operators, and that

$$\frac{i}{h} [\Delta_{f_1}, \Delta_{f_2}] = \Delta_{\{f_1,f_2\}} \quad .$$

One can readily pass back and forth between the two ways of describing this concept. See LMP, Vol. 2.

5.   THE SCHRÖDINGER QUANTIZATION RULES

Let (q^i) continue to denote a coordinate system for Q. The <u>Schrödinger rules</u> aim to assign skew-Hermitian operators to functions of (p_i, q^j):

For $f(q)$, Δ is multiplication by if.

For

$$f(p,q) = a^i(q)p_i \quad,$$

$$\Delta = a^i \frac{\partial}{\partial q^i} - \frac{1}{2} \frac{\partial a^i}{\partial q^j}$$

For

$$f = a^{ij}(q)p_i p_j \quad,$$

$$\Delta = i \frac{\partial}{\partial q^i} a^{ij} \frac{\partial}{\partial q^j} \tag{5.1}$$

This is a typical "symmetrization" rule: Write f in "symmetrized" form,

$$f = p_i a^{ij}(q)p_j \quad,$$

then make the assignments

$$p_i \rightarrow \frac{\partial}{\partial q^i}$$

$$a^{ij} \rightarrow \text{ multiplication by } ia^{ij} \tag{5.2}$$

$$p_j \rightarrow \frac{\partial}{\partial q^j}$$

Notice that (5.1) is obtained by making the substitutions (5.
for a^{ij} and p_i, and <u>interpreting the product as operator</u>
<u>product</u>.

Here is another way of looking at this. Assign:

$$a^{ij}(q)p_j \rightarrow a^{ij}\frac{\partial}{\partial q^j} - \frac{1}{2}\frac{\partial a^{ij}}{\partial q^j} \quad .$$

Assign to f the anticommutator of the operator assigned to
fp_i and the operator assigned to $a^{ij}p_j$. This is:

$$\frac{i}{2}\left(\frac{\partial}{\partial q^i}\left(a^{ij}\frac{\partial}{\partial q^j} - \frac{1}{2}\frac{\partial a^{ij}}{\partial q^j}\right)\right) + \left(a^{ij}\frac{\partial}{\partial q^j} - \frac{1}{2}\frac{\partial a^{ij}}{\partial q^j}\right)\frac{\partial}{\partial q^i} \quad ,$$

which is equal to

$$i\frac{\partial}{\partial q^i}a^{ij}\frac{\partial}{\partial q^j} + \frac{i}{2}\left(\frac{\partial a^{ij}}{\partial q^j}\frac{\partial}{\partial q^i} - \frac{1}{2}\frac{\partial}{\partial q^i}\left(\frac{\partial a^{ij}}{\partial q^j}\right) - \frac{1}{2}\frac{\partial a^{ij}}{\partial q^j}\frac{\partial}{\partial q^i}\right)$$

$$= i\frac{\partial}{\partial q^i}a^{ij}\frac{\partial}{\partial q^j} - \frac{i}{4}\frac{\partial^2(a^{ij})}{\partial q^i\partial q^j} \quad .$$

We see that this differs from the way we have "quantized" f
in formula (5.1) in the zero-th order term. This is, in fact
quite typical of the Schrödinger rules--the ambiguities in
ordering become insurmountable (in general) and one cannot

define a consistent quantization scheme for the case where $\underline{0}$
consists, say, of all polynomials in p and q.

Of course, the Schrödinger rules do suffice for the
systems usually encountered in (unconstrained) Newtonian
particle mechanics.

6. THE SCHRÖDINGER RULES FOR UNCONSTRAINED NEWTONIAN PARTICLE
    MECHANICS

In this case, we know that the configuration space Q is
a real vector space. Let

$$(x^i)$$

denote linear coordinates on Q. Let (p_i) denote the dual
linear coordinates on Q^d. Thus,

$$(x^i, p_j)$$

are linear coordinates for

$$Q \times Q^d \quad,$$

which can also be identified with

$$T^d(Q) \quad,$$

the phase space of the system.

In Newtonian particle mechanics, the Hamiltonian

$$H(x,p)$$

has the following form:

$$H = \frac{1}{2} m^{ij} p_i p_j + V(x) \tag{6.1}$$

The symmetric matrix

$$(m^{ij})$$

determines the <u>mass distribution</u>, while the function

$$x \rightarrow V(x)$$

on Q is the <u>potential function</u>.

The <u>classical equations of motion</u> are the Hamilton equations for the Hamiltonian H; namely,

$$\frac{dx^i}{dt} = \frac{\partial H}{\partial p^i} = m^{ij} p_j \tag{6.2}$$

$$\frac{dp_i}{dt} = -\frac{\partial H}{\partial x^i} = -\frac{\partial V}{\partial x^i} \tag{6.3}$$

We can put these equations into their usual "Newtonian" form by differentiating (6.2) with respect to t, and using (6.3):

$$\frac{d^2 x^i}{dt^2} = -m^{ij} \frac{\partial V}{\partial x^j} \quad ,$$

or

$$m_{ij} \frac{d^2 x^j}{dt^2} = -\frac{\partial V}{\partial x^j} \tag{6.4}$$

Notice that (6.4) reads:

> mass × accelleration = force.

One can also solve (6.2) for p in the form:

$$p_i(t) \;=\; m_{ij}\,\frac{dx^j}{dt} \tag{6.5}$$

(As usual in tensor analysis, (m_{ij}) is the inverse matrix to (m^{ij}).) This suggests that we should identify $p(t)$ with the linear momentum.

> momentum = mass × velocity.

To "quantize" such a system, we must describe a set $\underline{O}$ of functions of (x,p), forming the classical observables. They must form a Lie algebra under Poisson bracket, separate points on phase space, and contain the Hamiltonian function (6.1). In this case, the choice of $\underline{O}$ should be obvious:

$\underline{O}$ = set of functions $f(x,p)$ of the following form:

$$f(x,p) \;=\; a^{ij}p_i p_j + a^i_j x^j p_i + a^i p_i + a(x) \quad, \tag{6.6}$$

where the (a^{ij}, a^i_j, a^i) are constants, and $a(x)$ is an arbitrary function of x.

$\underline{O}$ may now be quantized by differential operators in the following way:

$$f \rightarrow ia^{ij} \frac{\partial^2}{\partial x^i \partial x^j} + a^i_j x^j \frac{\partial}{\partial x^i} - \frac{1}{2} a^i_i + a^i \frac{\partial}{\partial x^i} + ia(x)$$

$$(6.7)$$

In particular, the Hamiltonian H (i.e., the total energy) goes into the following operator:

$$H \rightarrow i \left(\frac{1}{2} m^{ij} \frac{\partial^2}{\partial x^i \partial x^j} + V(x) \right) \quad . \qquad (6.8)$$

In practice, the main problem is to find the spectrum of the operator on the right hand side of (6.8). Often, this is done by finding a unitary transformation on the Hilbert space in which the operator lives which sends it into a "known" operator. The analoque of this is, in classical mechanics, to find canonical transformation which sends the Hamiltonian into that of a "known" system.

Chapter 10

THE QUANTIZATION CONDITIONS OF BOHR,
SOMMERFELD AND EINSTEIN

1. INTRODUCTION

Tha famous "quantization" conditions developed by Bohr
and Sommerfeld before the full development of quantum
mechanics have always interested me, because of their evident
geometric and topological significance. In VB, Vol. 2, I
have already discussed the way they may most readily be
formulated in terms of modern mathematics. J. Keller and
V. Maslov have done extensive (and brilliant) work (well
before mine, of course) on the development of the underlying
"semi-classical approximation" to Schrödinger-like equations.
Recently, I have learned (from reading Lanczos' fascinating
summary [1] of Einstein's work in the period 1905-1915) that
Einstein already knew the interpretation I placed on the
rules in VB, Vol. 2. This explains the title of this chapter;
my intention is to review this material in more detail and to
develop various examples.

Here is the general set up. Q is a manifold, $T^d(Q)$
is its cotangent bundle, with the usual symplectic structure
defined by a closed two-form, ω. Let

$$H: T^d(Q) \to R$$

be a real-valued function. (Typically, H will be the
Hamiltonian of a mechanical system.)

   Let U be an open subset of Q, and

      S: U → R

a real-valued, C^∞ function on U. dS, its exterior derivative, is a field of covectors. In other words, it can be
identified with a cross-section map

      dS: U → $T^d(Q)$.

The following property holds:

$$(dS)^*(\omega) = 0 \qquad\qquad (1.1)$$

Definition. S is a solution of the Hamilton-Jacobi equation
with Hamiltonian H if there is a constant E ε R such that

$$(dS)^*(H) = E . \qquad\qquad (1.2)$$

   These are ideas which should be familiar (but in a
slightly different notation). They are typical of "classical
mechanics". It is well-known how solutions of (1.2) determine the solutions of the classical mechanics problem associated with the Hamiltonian H. Now, let us put a "quantum"
twist to them, in the following way.

2.   THE QUANTIZED HAMILTON-JACOBI EQUATION

Keep the notation outlined in Section 1. Let $\hbar$ be any real number. (Physically, it is the usual Planck's constant, divided by 2π.)

Let SO(2,R) be the group of rotations in R^2. Identify R^2 with C, the complex numbers, and identify SO(2,R) with the multiplicative group of complex numbers of absolute value one.

Definition.   A map

   g: Q → SO(2,R)

is a solution of the quantized Hamilton-Jacobi equation with quantum value E if it is continuous and if the following condition is satisfied:

> Q can be covered by open subsets U, in each
> of which there is a C^∞ function S: U → R
> which is a solution of the classical-mechanics
> Hamilton-Jacobi equation (1.2), such that, in U,

$$g = \exp(iS/\hbar) \qquad (2.1)$$

Remark.   Here is what is involved intuitively. (Judging by Lanczos' remarks [1], these ideas were well understood by Einstein before 1920, but apparently not necessarily by his

contemporaries.) As we know, "solving" the classical mechanics
problem involves finding a function S on open subsets of Q
which solve the Hamilton-Jacobi equation (1.2). In typical
examples, these functions will not be globally defined on Q,
but will have singularities, and will be "multi-valued".
Requiring that such a continuous, globally-defined g exists
restricts the "discontinuities" possible in the solution of
the H-J equation; namely, they must be of the form:

$$2\pi\hbar \times \text{integer}$$

A value of E for which there exists such a global function is said to be a quantized value of H. (The corresponding idea in the usual Hilbert space formulation of quantum
mechanics is that E lies in the spectrum (in the Hilbert
space sense) of the self-adjoint operator corresponding to H.)
To a mathematician, this is an ideal formulation of the
physical idea of "semi-classical quantization", since it is
general, coordinate-free, and provides interesting mathematical concepts to think about. (For example, there are
questions that one trained in topology would almost immediately
begin to ask himself.) Instead of this approach, one finds
the idea expounded in the standard physics literature in an
incredibly mixed up and complicated way, involving all sorts
of special jargon ("action-angle variables", "adiabatic

invariants", etc.). While these concepts underlying this jargon may, in fact, be interesting, in their own right, they do not help one to understand the underlying idea in a simple and natural mathematical way. They also have the unfortunate aesthetic defect of being so dependent on choice of coordinates!

3.   SYSTEMS WITH ONE DEGREE OF FREEDOM

Suppose that Q is one-dimensional. Let q be a local coordinate for Q. Let

$$(q,p)$$

be the corresponding coordinate for $T^d(Q)$; i.e., if

$$\theta \;=\; a\,dq \quad,$$

then

$$q(\theta) \;=\; q$$
$$p(\theta) \;=\; a$$

(Thus, p is essentially $\partial/\partial q$.)

Let $H(q,p)$ be the Hamiltonian. The Hamilton-Jacobi equation is

$$H\left(q, \frac{dS}{dq}\right) \;=\; E \quad. \tag{3.1}$$

We must find the conditions that there exist a map

$$g: Q \rightarrow SO(2,R) \quad,$$

such that, in the local coordinate q,

$$g(q) = e^{iS(q)/\hbar} \tag{3.2}$$

Example. The harmonic oscillator.

$$h(q,p) = q^2 + p^2 , \qquad\qquad -\infty < p,q < \infty . \tag{3.3}$$

(3.1) then takes the form:

$$q^2 + \left(\frac{dS}{dq}\right)^2 = E ,$$

or

$$\frac{dS}{dq} = \sqrt{E - q^2} .$$

$$S = \int \sqrt{E - q^2} \, dq$$

$$= \quad , \text{ substituting } q = \sqrt{E} \sin \alpha$$

$$(E) \int \cos^2 \alpha \, d\theta$$

$$= (E) \int \frac{1+\cos^2 \alpha}{2} \, d\theta$$

$$= \frac{1}{2} \theta + \frac{\sin^2 \alpha}{4} \quad (E)$$

or

$$g = c e^{i/\hbar(\alpha^2 + \sin 2\alpha)E} \tag{3.4}$$

Now, α can serve as well as g as a <u>local</u> coordinate for Q. However, a <u>necessary</u> condition that g be defined on $-\infty < g < \infty$ is that

$$g\big|_{\alpha=\pi} = g\big|_{\alpha=0} \; .$$

or,

$$e^{Ei/\hbar(\pi/2)} = 1 \; ,$$

or

$$\cos\left(\frac{E}{\hbar}\frac{\pi}{2}\right) = 1 \; ,$$

$$\sin\left(\frac{E}{\hbar}\frac{\pi}{2}\right) = 0 \; ,$$

or

$$E = 2n\hbar \; , \tag{3.5}$$

for some integer n.

This is, of course, the usual "quantization condition" for the harmonic oscillator.

<u>Exercise</u>. Show that condition (3.5), with integer $n > 0$, is a sufficient condition for the existence of a C^∞ map $g: Q \to SO(2,R)$ which is locally of the form (3.4).

Return to the case of a general one-degree of freedom mechanical system defined by a Hamiltonian function

$$H: T^d(Q) \rightarrow R \quad ,$$

$$\dim Q = 1 \quad .$$

Sommerfeld showed how to describe the quantization condition in a completely intrinsic way in this case.

Let θ be the contact one-form on $T^d(Q)$. Then, in the local coordinates (q,p) arising from a local coordinate (q) for θ,

$$\theta = p \, dq \quad .$$

For $E \varepsilon R$, set

$$
\begin{aligned}
M_E = \; & \text{set of points of } T^d(Q) \text{ at} \\
& \text{which } H = E \quad .
\end{aligned}
\tag{3.6}
$$

Theorem 3.1. Suppose that E is a value of H such that:

   a) E is a regular value of H, in the sense that

$$dH \neq 0$$

   at all points of M_E.

   b) M_E is compact and connected.

Then, E is quantized according to the semi-classical quantization rules if and only if

$$
\boxed{\int_{M_E} \theta = i2\pi n \hbar}
\tag{3.7}
$$

for some integer n.

Proof. Hypothesis a) guarantees that M_E is a sub-manifold of $T^d(Q)$.

Let (q,p) be the usual coordinates for $T^d(Q)$. Suppose that

$$g: Q \rightarrow SO(2,R)$$

is a continuous map which satisfies the quantized Hamilton-Jacobi equation. Recall that this means that, locally, there is a solution $S(q)$ of the "classical" H-J equation such that:

$$g(q) = e^{iS(q)/\hbar} \qquad (3.9)$$

Then,

$$dgg^{-1} = \frac{i}{\hbar} dS \qquad (3.10)$$

Hence,

$$\eta \equiv \frac{\hbar}{i} dgg^{-1} = , \text{ locally, } dS \qquad (3.11)$$

is a real, closed, globally defined one-form on Q. In other words, η is a cross-section map

$$Q \rightarrow T^d(Q) \quad ,$$

and

$$\eta^*(d\theta) = 0 \qquad (3.12)$$

That S solves the H-J equation, means that

$$\eta^*(H) = E \quad , \qquad (3.13)$$

i.e.,

$$\eta(Q) \subset M_E \tag{3.14}$$

Now,

$$\pi^*(\eta) = \text{, locally, } \frac{\partial S}{\partial q} dq \quad . \tag{3.15}$$

Lemma.

$$\boxed{p = \frac{\partial S}{\partial q} \text{ on the submanifold } M_E \text{ of } T^d(Q).} \tag{3.16}$$

Proof. We know that M_E is one dimensional. It is filled up by solutions of the Hamilton equation:

$$\frac{dq}{dt} = \frac{\partial H}{\partial p}$$
$$\tag{3.17}$$
$$\frac{dp}{dt} = -\frac{\partial H}{\partial q}$$

such that also:

$$H(q(t), p(t)) = E \quad .$$

Now, we can also solve these equations by solving the H-J equation for a function $S(q)$, then solving the equations

$$\frac{dq}{dt} = \frac{\partial H}{\partial p}\left(q(t), \frac{\partial S}{\partial q}(q(t))\right)$$
$$\tag{3.18}$$
$$p(t) = \frac{\partial S}{\partial q}(q(t))$$

By <u>uniqueness</u> of solutions of ordinary differential equations,
the solutions of (3.16) must be solutions of (3.17), i.e.,
$p(t)$ must satisfy the <u>second</u> equation of (3.18). But, this
is exactly what is involved in (3.16).

Relations (3.15) and (3.16) are (from a general topological differential geometric point of view) the key relations.
For they imply that:

$$
\boxed{
\begin{aligned}
\pi^*(\eta) \quad & \text{restricted to} \quad M_E \\
= \quad \theta \quad & \text{restricted to} \quad M_E.
\end{aligned}
}
\tag{3.19}
$$

As a consequence of (3.19), we have:

$$
\int_{M_E} \theta \;=\; \int_{M_E} \pi^*(\eta)
$$

$$
= \quad , \text{ using } (3.11),
$$

$$
\frac{\hbar}{i} \int_{M_E} d\pi^*(g)\,\pi^*(g^{-1})
$$

$$
= \quad , \text{ locally,}
$$

$$
\frac{\hbar}{i} \int_{M_E} d \log (\pi^*(g))
\tag{3.20}
$$

Of course, "log $\pi^*(g)$" is a multi-valued function on M_E. When the integral on the right hand side of (3.20) is evaluated by using local coordinates for M_E, it is seen that the right hand side of (3.20) is

$$2\pi \frac{\hbar}{i} n \quad , \tag{3.21}$$

for some integer n. (Another way of putting this is that the cohomology class of M_E represented by the closed differential form involved in the integrand in (3.20) is an <u>integer multiple</u> of the generating class of $H^1(M_E,Z)$.) This proves relation (3.7).

<u>Exercise</u>. Finish the proof of Theorem 3.1 by proving the converse, i.e., that 3.7 implies that there is a global, continuous solution of the "quantized H-J equation". Is there a C^∞ solution?

<u>Remark</u>. I do not recall seeing this form of the argument anywhere else, and it might even be new. It suggests various generalizations. Here are obvious questions:

What happens in case θ is higher dimensional?

What happens in case $SO(2,R)$ is replaced by another compact Lie group, or, more generally, by a compact manifold?

I believe this latter question would be particularly interesting physically, for example, in relation to the question of
the "semi-classical" approximation of the Dirac equation.
In fact, since the Dirac equation is a generalization of the
Schrödinger equation, there should be a generalization of the
H-J equation which stands in the same relation to the Dirac
equation as the H-J equation itself stands to the Schrödinger.

Another interesting question is:

> What is the generalization if M_E
> is non-compact?

Notice that we have not really used compactness of M_E in this
argument in a very strong way. If we interpret

$$\int_{M_E} \theta$$

in some "generalized function" sense (e.g., in terms of some
limiting relation), some analogue of (3.7) should hold. (Of
course, the possibility that "$n = \infty$" should be left open
also!) Perhaps Maslow's work [1] on the relation to the
spectral theory (in the Hilbert space sense) of the Schrödinger
operator will give some insights into the correct way of
interpreting the conditions.

The connections with topology are also interesting. The integer n is essentially the degree of the mapping

$$g: M_E \rightarrow SO(2,R)$$

between compact orientable manifolds. Let us recall what is involved here.

Suppose that N,N' are compact, oriented manifolds of some dimension, and that

$$\phi: N \rightarrow N'$$

is a C^∞ map between them. Let dp,dp' be positively oriented volume element forms on N and N'. Let f be the function such that:

$$\phi^*(dp') = j\, dp \quad .$$

Then, the integral

$$\int_M j(p)\, dp \tag{3.22}$$

is called the degree of the mapping. It is always an integer.

Here is a typical situation where one can calculate the degree in a more geometric way.

The points p ε M such that

$$j(p) \neq 0$$

are called the regular points of the map ϕ. (In local coordinates, j is essentially the Jacobian determinant of the map ϕ.) The points p' ε M' such that

$$\phi^{-1}(p')$$

consists only of regular points are called the <u>regular values</u>
of ϕ. A general theorem of Sard asserts that the set of non-
regular values is a set of <u>measure zero in</u> M!

Frequently, we have the following condition satisfied:

> The number of points in $\phi^{-1}(p')$ is
> finite and constant as p' runs through
> the set of regular values of ϕ.

In this case, the number of points in the "generic" fiber
$\phi^{-1}(p')$ is <u>equal</u> to the degree of the mapping, as defined by
3.21. This number can be thought of as the <u>ramification number</u>
of ϕ or "degrees of multi-valuedness" of ϕ^{-1}. The key fact
that quantum conditions can be expressed in terms of "degrees
of mappings" is of great interest to the topologist. This
fact also comes to the foreground in Maslow's general "WKB"
theory.

4.   A GENERALIZATION OF THE BOHR-SOMMERFELD QUANTIZATION
        RULES FOR SYSTEMS WITH MANY DEGREES OF FREEDOM

        We now consider generalizations of the method of proof
given in Section 2 of the Bohr-Sommerfeld quantization rules
for systems of one degree of freedom.

        Let Q be a manifold of an arbitrary number of dimen-
sions. Let

$$M \;=\; T^d(Q)$$

be its cotangent bundle; let

$$H: M \rightarrow R$$

be a Hamiltonian function; and for $E \;\epsilon\; R$, M_E be the subset

$$H^{-1}(E)$$

of M. Suppose that E is a regular value of H, i.e., that

$$\boxed{M_E \;\; \text{is a codimension one submanifold of} \;\; M}$$

Let θ be the contact one-form on $M = T^d(Q)$.
        Given a map

$$g: Q \rightarrow SO(2,R) \quad ,$$

set

$$\eta \;=\; \frac{\hbar}{i} \, dg g^{-1} \quad . \tag{4.1}$$

η is a closed one-form on Q.

Recall that g defines a solution of the <u>quantized H-J equation</u> with E <u>as the quantized energy value</u>, if:

$$\eta(Q) \subset M_E \qquad\qquad (4.2)$$

<u>Theorem 4.1</u>. Suppose that (4.1) and (4.2) are satisfied. Let σ be a compact, one-dimensional submanifold of M_E. Then,

$$\int_\sigma \eta = 2\pi n \hbar \quad , \qquad\qquad (4.3)$$

for some integer n.

<u>Corollary to Theorem 4.1</u>. Suppose that:

$$\eta \text{ restricted to } \sigma = \theta \text{ restricted to } \sigma \qquad (4.4)$$

Then,

$$\int_\sigma \theta = 2\pi n \hbar \qquad\qquad (4.5)$$

for some integer n.

<u>Remark</u>. Notice that the difference between conditions (4.4) and (4.5) is that the form η depends on the choice of the map g, while θ is intrinsically associated to Q. In this way we can hope to obtain necessary conditions of a global

nature for the existence of such a map g, i.e., for the
"quantization" of the energy value E.

Now, the obvious candidates for submanifolds σ to whi
to apply condition (4.5) are the <u>periodic orbits</u> of the Hamiltonian system. We shall now briefly study their properties.

5.   QUANTIZATION CONDITIONS FOR THE PERIODIC ORBITS

Continue with the notation of Section 4. Let A_H be
the vector field on $T^d(Q) \equiv M$ such that:

$$dH = A_H \lrcorner \omega \qquad (5.1)$$

In canonical coordinates (p_i, q^i),

$$A_H = \frac{\partial H}{\partial p_i} \frac{\partial}{\partial q^i} - \frac{\partial H}{\partial q^i} \frac{\partial}{\partial p_i} \qquad (5.2)$$

We see that the <u>orbit curves</u> of A_H are the <u>solutions of the</u>
<u>Hamiltonian equations with Hamiltonians</u> H.

Further,

$$A_H(H) = \{H,H\} = 0 \qquad (5.3)$$

Of course, this is just the familiar fact that the Hamiltonian
(which is often the <u>total energy</u> of the system) is <u>conserved</u>.
In our situation, it implies that:

A_H is tangent to each submanifold
$$M_E \equiv H^{-1}(E) \quad .$$

Let us suppose that A_H generates a <u>global</u> one-parameter group of diffeomorphisms of M_E. (By general Lie theory, it always generates a "local" group. If M_E is compact, which is a typical situation in the applications, A_H will indeed generate such a global group.) This means that the additive group R of the real line acts on M_E; the orbits are the <u>orbit curves</u> of A_H, i.e., the solutions of the Hamilton equations. Let

$$t \rightarrow \sigma(t) \quad , \qquad\qquad -\infty < t < \infty$$

denote such an orbit. It is said to be <u>periodic</u> if there is a number $T > 0$ such that

$$\sigma(t+T) = \sigma(t) \qquad\qquad\qquad (5.3)$$

for all $t \in R$.

<u>Exercise.</u> If (5.3) holds for one value, say t_0, show that it holds for all values of t.

<u>Exercise.</u> If (5.3) holds for one value of T, show that there is a value $T_0 > 0$ such that all other values are integer multiples of T_0. This value is called the <u>period</u> of the periodic orbit.

We see then that a periodic orbit determines a mapping, which we also denote by σ, of the real numbers R <u>modulo</u> T_0.

This is a submanifold mapping. (Recall that a submanifold
mapping is a mapping $\phi: N \rightarrow M$ between manifolds whose differential $\phi_*: T(N) \rightarrow T(M)$ is one-one.)

We denote this submanifold mapping by σ.
(If the notation were too ambiguous, we
would use σ_{T_0} .)

Thus, if θ is a one-form on   M,   and if σ is the
submanifold corresponding to a periodic orbit,

$$\int_\sigma \theta \qquad\qquad (5.4)$$

can be defined as usual.

Now, let θ be the contact one-form on $T^d(Q) \equiv M$.
In the physics literature (e.g., Goldstein [1], Corben-Stehle
[1]) (5.4) is then called the action over the periodic orbit.
I find this a confusing terminology (because of confusion with
the "action" as the integral of the Lagrangian), hence, will
not use it.

If

$$t \rightarrow (q(t), p(t))$$

are the coordinates of σ,   then (5.4) takes the form:

$$\int_{0}^{T_0} p_i(t) \frac{dq^i}{dt}(t) \, dt \qquad (5.5)$$

Now, suppose that

$$g: Q \to SO(2,R)$$

is a solution of the <u>quantized Hamilton-Jacobi equation</u>. Recall that this means that:

Each point of Q has a solution S of the Hamilton-Jacobi equation defined in an open subset about it such that

$$g = e^{iS/\hbar} \qquad (5.6)$$

<u>Remark</u>. This condition might have to be weakened. Perhaps, one might require that g be <u>continuous</u>, as a map between topological space, and that there be an open, dense subset Q' of Q such that (5.6) holds in open subsets of Q'.

As before, define η as a closed one-form on Q associated with g. Namely,

$$\eta = \frac{\hbar}{i} dg g^{-1} = \quad , \text{ locally,} \quad dS \quad . \qquad (5.7)$$

Regard η also as a cross-section map

$$\eta: Q \to T^d(Q) \equiv M \qquad\qquad (5.8)$$

Recall that the condition that S is a solution of the H-J equation is that

$$\eta(Q) \subset M_E \quad . \qquad\qquad (5.9)$$

Theorem 4.1. There is a vector field B_g on Q such that:

$$\eta_*(B_g) = A_H \quad , \qquad\qquad (5.10)$$

i.e., the image under η of the orbit curves of B_g are orbit curves of A_H. In local coordinates (q^i)

$$B_g = \frac{\partial H}{\partial p_i}\left(q, \frac{\partial S}{\partial q}\right)\frac{\partial}{\partial q^i} \qquad\qquad (5.11)$$

Proof. It is well known (see DGCV, for example), that, for each solution S of the H-J equation

$$H\left(q, \frac{\partial S}{\partial q}\right) = E \quad ,$$

the solutions of the following equations in Q-space are also solutions of the Hamilton-Jacobi equations in (P,Q)-space:

$$\frac{dq^i}{dt} = \frac{\partial H}{\partial p^i}\left(q(t), \frac{\partial S}{\partial q}(q(t))\right) \qquad\qquad (5.12)$$

Exercise. Prove this by a direct calculation.

The statement of Theorem 5.1 is nothing but a restatement of these facts

<u>Remark</u>. Here is another way of proving Theorem 5.1 in a completely coordinate-free way.

Consider

$$\eta(Q) \subset M_E \subset M \equiv T^d(Q) \quad .$$

$\eta(Q)$ is a submanifold of M_E, and η is a diffeomorphism between it and Q. One proves that:

$$\boxed{A_H \text{ is tangent to the submanifold } \eta(Q)}$$

B_g is then the vector field given by the following formula on Q:

$$B_g = \eta_*^{-1}(A_H \text{ restricted to } A_H)$$

Here is a diagram

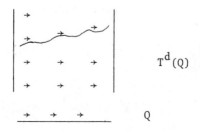

The arrows represent the Hamiltonian vector field A_H.

In terms of the <u>calculus of variations</u>, this phenomeno$\;$ is closely related to the <u>extremal field</u> concept. Another classical way of thinking about it is to identify the Hamilto$\;$ equations as the <u>characteristic equations</u> of the Hamilton-Jacobi <u>partial</u> differential equation.

We can now return to the quantization conditions.

<u>Theorem 4.2</u>. Let σ be a periodic orbit of A_H which is th$\;$ image under η of an orbit of the vector field B_g. Let θ be the contact one-form on $M \equiv T^d(Q)$. Then,

$$\int_\sigma \theta = 2\pi n\hbar \qquad (5.1$$

for some integer n .

<u>Proof</u>. We use local coordinate

$$(q^i, p_i) \equiv (q, p)$$

Let $t \to (q(t), p(t))$ be coordinates for σ. Then,

$$\frac{dq^i}{dt} = \frac{\partial H}{\partial p_i}\left(q(t), \frac{\partial S}{\partial q}(q(t))\right)$$

$$\qquad (5.1$$

$$p_i(t) = \frac{\partial S}{\partial q^i}(q(t))$$

Hence, on σ,

$$\theta = p_i dq^i$$

$$= \text{, using (4.14),}$$

$$\frac{\partial S}{\partial q^i} dq^i$$

$$= \eta \quad .$$

Thus, $\theta = \eta$ on σ, and the Corollary to Theorem 4.1 implies (5.13).

6.   HAMILTONIANS WHICH ADMIT SEPARABLE COORDINATE SYSTEMS

In previous sections, we have derived necessary conditions that a particular value E of the energy be "quantized",
according to the Bohr-Sommerfeld-Einstein rules. Now, we turn
to finding sufficient conditions.

The only systems for which one knows very much in a
global way about solutions of the Hamilton-Jacobi equation are
those which are called separable in the classical literature,
i.e., whose Hamilton-Jacobi equation can be "solved" by means
of the classical method of "separation of variables". In this
case, solutions of the multi-degrees-of-freedom H-J equation
can be built up from solutions of one degree of freedom equations.

Continue with the notation of previous sections.   Let

$$1 \leq i,j \leq n \equiv \dim Q \quad .$$

Adopt the summation convention on these indices.   Let

$$(q^i)$$

be a <u>coordinate system</u> for an open subset of Q.   Let

$$(q^i, p_i)$$

be the corresponding coordinate system for that part of $T^d(Q)$
above the open subset.   Then, the contact structure on $T^d(Q)$
is defined in these coordinates by

$$\theta = p_i dq^i \quad ,$$

and the symplectic structure by

$$\omega = dp_i \wedge dq^i \quad .$$

Let $H: T^d(Q) \rightarrow R$ be a Hamiltonian function.   In these variables, H is a function

$$H(q,p) \quad .$$

Consider a space Y of dimension $(n-1)$, with variab.

$$y = (y_a) \quad , \qquad\qquad 1 \leq a,b \leq n-1 \quad .$$

Also, fix a value

$$E$$

of the <u>total energy</u> of the system.

Definition. H is said to be separable with respect to the coordinate system (q^i) if there are functions

$$W_1(q^1;y),\ldots,W_n(q^n;y)$$

on $R \times R^{n-1}$ such that the sum function

$$W(q;y) = W_1(q^1;y) + \cdots + W_n(q^n;y) \qquad (6.1)$$

satisfies the following conditions:

a) $H\left(q, \dfrac{\partial W}{\partial q}\right) = E$,

   i.e., for fixed g, the function

   $g \rightarrow W(q,y)$

   is a solution of the Hamilton-Jacobi equation
   with total energy E.

b) Rank of the matrix

$$\left(\dfrac{\partial^2 W}{\partial q^i \partial y_q}\right) \qquad (6.2)$$

   is always n-1.

Let M_E be the subset of $T^d(Q) \equiv M$ on which H has the energy E. By condition (6.2) the map

$$\phi: (q,p) \qquad \left(q, \ p_i = \frac{\partial W}{\partial q^i}\right) \qquad (6.3)$$

is a _submanifold mapping_. In particular, its image _locally_
fills up M_E, i.e., the map (6.3) provides a _parameterization_
of the submanifold M_E of M.

Now, because of (6.1), the map (6.3) takes the followin
form:

$$p_1 = \frac{\partial W_1}{\partial q^1} (q^1, y)$$

$$\vdots \qquad\qquad\qquad (6.4)$$

$$p_n = \frac{\partial W_n}{\partial q^n} (q^n, y)$$

Let

$$\theta_E = \theta \text{ restricted to the submanifold } M_E \quad (6.5)$$

Then, by (6.4), we have:

$$\theta = p_1(q^1; y) dq^i + \cdots + p_n(q^n; y) dq^n \qquad (6.6)$$

Now, for each $y \in Y$, there is a vector field

$$A_y$$

on Q defined by the following formula:

$$A_y = \frac{\partial H}{\partial p_i} \left(q, \frac{\partial W}{\partial q} \right) \frac{\partial}{\partial q^i} \qquad (6.7)$$

Since, for fixed y, the function $q \to W(q,y)$ is a solution of the H-J equation, we see that A_y is a vector field on Q whose orbits are projections of orbits of A_H, i.e., of solutions of Hamilton equations.

Let us suppose that globally Q is the product

$$Q = Q_1 \times \cdots \times Q_n$$

of one-dimensional manifolds, for which the coordinates are $q^1,\ldots,q^n$, (In fact, this will not be precisely satisfied in the interesting examples, but the techniques can be modified.) For fixed $y \in Y$, define one-forms

$$\theta_y^1,\ldots,\theta_y^n$$

on $Q_1,\ldots,Q_n$, by the following formula:

$$\theta_y^1 = p_1(q^1,y)\, dq^1$$
$$\vdots \qquad\qquad\qquad (6.8)$$
$$\theta_y^n = p_n(q^n,y)\, dq^n \quad,$$

where the p's are defined by (6.4). Then, (6.6) can be written as follows:

$$\theta = \theta^1 + \cdots + \theta^n \qquad\qquad (6.9)$$

Here is the main result:

Theorem 6.1.   Let y be an element of Y such that an orbit of A_s is periodic. Let $\sigma_1,\ldots,\sigma_n$ be curves in $Q_1,\ldots,Q_n$

such that, <u>in terms of the</u> (q^i) <u>coordinates</u>,

$$\sigma = \sigma_1 \times \cdots \times \sigma_n \quad .$$

Suppose that $j_1, \ldots, j_n$ are integers such that:

$$\int_1 \theta_y^1 = 2\pi\hbar j_1$$

$$\vdots$$

$$\int_n \theta_y^n = 2\pi\hbar j_n$$

(6.1

Then, there is a

$$g: Q \to SO(2,R)$$

which is a <u>global</u> solution of the <u>quantized</u> Hamilton-Jacobi equation, such that, in the (q^i)-coordinates:

$$g(q) = e^{iW(q;y)/\hbar}$$

(6.1

Remark. Conditions 6.10 are the <u>Bohr-Sommerfeld quantization rules</u>. They can be interpreted physically by saying that the "action" variables

$$J_1 = \oint \theta_1$$

$$\vdots$$

$$J_n = \oint \theta_n$$

take on "quantized" values. See Goldstein [1] and Born [1]
for an explanation.

   Proof of Theorem 6.1. The differential equations of the
orbits of A_s are given as follows:

$$\frac{dq^1}{dt} = \frac{\partial H}{\partial p_1} (q(t), p_1(q^1(t),y))$$

$$\vdots \tag{6.12}$$

$$\frac{dq^n}{dt} = \frac{\partial H}{\partial p_n} (q(t), p_n(q^n(t),y))$$

Further,

$$p_1 = \frac{\partial W_1}{\partial q^1} (q^1,y)$$

$$\vdots$$

$$p_n = \frac{\partial W_n}{\partial q^n} (q^n,y)$$

We see that condition (6.10) means that

$$q_1(q^1) = e^{iW_1(q^1,y)/\hbar}$$

$$\vdots$$

$$q_n(q^n) = e^{iW_n(q^n,y)/\hbar}$$

are globally defined on $Q_1,\ldots,Q_n$.

Final Remarks. There are many things I do not understand ab
this topic. For example, in the separable case, are there a
other quantized values of E than those determined as expla
in Theorem 5.1 by the Bohr-Sommerfeld conditions? Is there
sharper, more definitive way of stating the Einstein interpretation of these equations, i.e., the existence of a globa
solution of the quantized H-J equation?

Here is one clue to such an ultimate, mathematically
correct theory. In the physics literature, there are vague
references to Ehrenfest's Adiabatic Hypothesis as providing
the correct physical explanation of which variables are to b
"quantized". (See Sommerfeld's book "Atomic structure and
spectral lines", Chapter V, Section 7.) Now, it is clear to
me that there is some relation of the physicist's notion of
adiabatic invariant and the concept of integral invariant du
to Poincaré and E. Cartan. (See Cartan's book "Leçons sur 1
invariants intégraux".) Thus, I believe that there should b
some way of discussing the quantization of integral invarian

7.   THE KEPLER PROBLEM

Here,

$$\dim Q = 3 \quad .$$

Label the variables (q) as:

$$(r,\theta,\phi) \ ,$$

(spherical coordinates in R^3, of course). Label the dual variables as

$$(p_r,p_\theta,p_\phi) \ \ .$$

Then,

$$H \ = \ \frac{1}{2}\left(p_r^2 + \frac{p_\theta^2}{r^2} + \frac{p_\phi^2}{r^2 \sin^2 \theta}\right) \ - \ \frac{k}{r} \tag{7.1}$$

(I am following the notation used by Goldstein [1], Chapter 9, Section 7.)

Let us look for a solution of the Hamilton-Jacobi equation of the following form:

$$W \ = \ W_r(r) + W_\theta(\theta) + W_\phi(\phi) \tag{7.2}$$

(W implicitly depends on two additional variables (y_1,y_2) also.)

The first relation is:

$$\frac{\partial W_\phi}{\partial \phi} \ = \ y_1$$

or

$$W_\phi \ = \ y_1\phi \tag{7.3}$$

Substituting back into the H-J equation, it leads to:

$$\left(\frac{\partial W_\theta}{\partial \theta}\right)^2 \ + \ \frac{y_1^2}{\sin^2 \theta} \ = \ y_2^2 \tag{7.4}$$

$$\left(\frac{\partial W_r}{\partial r}\right)^2 + \frac{y_2^2}{r^2} = 2(E+(k/r)) \tag{7.5}$$

$$W_\theta = \int \sqrt{y_2^2 - \frac{y_1^2}{\sin^2 \theta}} \; d\theta \tag{7.6}$$

$$W_r = \sqrt{2(E+(K/R)) - \frac{y_2^2}{r^2}} \; dr \tag{7.7}$$

Hence,

$$p_\phi = y_1 \tag{7.8}$$

$$p_\theta = \sqrt{y_2^2 - \frac{y_1^2}{\sin^2 \theta}} \tag{7.9}$$

$$p_r = \sqrt{2(E+(k/r)) - \frac{y_2^2}{r^2}} \tag{7.1}$$

The orbits of A_y are solutions of the following
equation:

$$\frac{d\phi}{dt} = \frac{y_1^2}{r^2 \sin^2 \theta} \tag{7.11}$$

$$\frac{d\theta}{dt} = \frac{p_\theta}{r^2} = \frac{1}{r^2} \left(\sqrt{y_2^2 - \frac{y_1^2}{\sin^2 \theta}} \right) \tag{7.12}$$

$$\frac{dr}{dt} = p_r = \sqrt{2(3+(k/r)) - \frac{y_2^2}{r^2}} \tag{7.13}$$

It is well known that (if $k > 0$) the solutions of these equations are indeed periodic.

Set:

$$J_\phi(y) = \oint p_\phi d\phi = \oint y_1 \, d\phi$$

$$= 2\pi y_1 \tag{7.14}$$

$$J_\theta(y) = \oint p_\theta d\theta = \oint \sqrt{y_2^2 - \frac{y_1^2}{\sin^2 \theta}} \, d\theta \tag{7.15}$$

$$J_r(y) = \oint p_r \, dr = \oint \sqrt{2(E+(k/r)) - \frac{y^2}{r^2}} \qquad (6.16$$

Here the symbols $\oint$ indicate that we integrate over the interval of the variable corresponding to a complete <u>period</u> of the orbit of A_y. These integrals are evaluated by Goldstein [1], Chapter 9, Section 7. The important relation is the following one:

$$E = -\frac{\text{constant}}{(J_r + J_\theta + J_\phi)^2} \qquad (6.17$$

When this relation is <u>quantized</u>, i.e., the point y is chose so that J_r, J_θ, J_ϕ are <u>integer multiples of</u> $2\pi i\hbar$, (6.17) gives the usual formulas for the <u>energy spectrum of the hydro</u><u>gen atom</u>.

We shall further consider the differential geometric meaning of formulas like (6.17) in a later volume. Note also that there should be a relation to the group-theoretic inter pretation of the formula for the energy spectrum of the hydrogem atom-Kepler problem (See VB, Vol. 2, and IM, Vol.

Chapter 11

CANONICAL TRANSFORMATIONS GENERATED BY
COMPLETE SOLUTIONS OF THE HAMILTON-JACOBI EQUATION

1.   INTRODUCTION

In the classical work on mechanics, one finds various
general methods for "solving" the equations of motion of a
mechanical system. The method of Jacobi is indirect (since it
depends on finding a family of solutions of a <u>partial</u> differential equation), but very useful when it applies. Historically,
it has, in fact, been found to apply to many sorts of problems,
and has strongly influenced the development of celestial mechanics and quantum mechanics.

In this chapter I will briefly describe the underlying
general differential geometric theory, then work out two main
examples, the <u>harmonic oscillator</u> and <u>hydrogen atom-Kepler</u>
<u>problem</u>.

2.   SYMPLECTIC MANIFOLDS AND CANONICAL COORDINATES

<u>Definition</u>. A manifold M is said to have a <u>symplectic</u>
<u>structure</u> if it is even dimensional, and if it has a closed
two-form ω defined on it which is of maximal rank at each
point.

<u>Remark</u>. We say, for short, that M is a <u>symplectic manifold</u>.
I sometimes also use the term "canonical manifold", leaving
the name "symplectic" free to denote a manifold with an arbitrary closed two-form. However, the terminology used above seems
to be more standard in the current literature.

Suppose M has such a structure. ω is called the
<u>symplectic form</u>. If:

$$\dim M = m = 2n$$

then the "maximal rank" condition means that the 2n-form

$$\omega \equiv \underbrace{\omega \wedge \ldots \wedge \omega}_{n \text{ factors}} \qquad (2.1)$$

is non-zero at each point of M. In particular, M is an
<u>orientable</u> manifold, and ω^n can serve as a <u>volume element
form</u>, with which one can build up an <u>integration theory</u>. (See
DGCV.)

Introduce local coordinates. Choose the following range
of indices:

$$1 \leq i,j \leq n \qquad (\equiv 1/2 \text{ dimension of } M)$$

<u>Definition</u>. A set of functions on M

$$(y_i, x^i)$$

is said to form a <u>canonical coordinate system</u> for M if:

$$\omega = dy_i \wedge dx^i \tag{2.2}$$

<u>Theorem 2.1</u>. Such a set of functions are functionally indepen-
dent, hence indeed form a local coordinate system for M. Also,

$$\omega^n = dy_1 \wedge \ldots \wedge dy_n \wedge dx^1 \wedge \ldots \wedge dx^n \tag{2.3}$$

i.e., the volume element form associated to the symplectic
structure is the coordinate-system volume element.

> <u>Proof</u>. First, one proves (2.3) by straight forward
algebra, which is left to the reader. (It basically amounts to
showing that an n×n symplectic matrix has determinant one.)

Once proven, (2.3) shows that the $dy_1, \ldots, dx^n$ are
linearly independent, hence that the $y_1, \ldots, x^n$ are functionally independent. Since there are $2n = m = \dim M$ such
functions, their differentials form a basis for the cotangent
space at each point of M, which implies (by the Implicit
Function Theorem) that they form a coordinate system.

It is a well-known classical result (ascribed to Darboux
in recent literature, but proved even earlier) that such canonical coordinate systems exist <u>locally</u>. See DGCV and Cartan's
"Invariants Integraux". The more classical approach is
presented in Caratheodory [1].

Definition. A submanifold map

$$\alpha: N \to M$$

is said to define a _Lagrangian submanifold_ (of the symplectic
structure on M) if the following conditions are satisfied:

$$\alpha^*(\omega) = 0 \qquad\qquad\qquad (2.4)$$

$$\dim N = n = \frac{1}{2} \dim M$$

Remark. Cartan would say that α is a _maximal integral submanifold of_ ω, (or, rather, of the ideal of forms generated
by ω). I am not particularly happy with the use of Lagrange's
name here (I do not know the history), but this terminology
seems to be now standard.

Definition. A submanifold map

$$\alpha: N \to M \quad,$$

with $\dim N = 1/2 \dim M = n$, is said to be _transversal to the_
canonical coordinate system (x_i, y^i) if:

$$\alpha^*(dx^1 \wedge \ldots \wedge dx^n) \neq 0 \qquad\qquad (2.5)$$

Condition (2.5) means that the functions $\alpha^*(x^1), \ldots, \alpha^*(x^n)$
form a local coordinate system for N.

Suppose now that, in addition to (2.5), α is a
Lagrangian submanifold mapping. Now, using (2.2),

$$\omega = d(y_i dx^i) \quad .$$

Hence, using (2.4),

$$d(\alpha^*(y_i)\, d(\alpha^*(x^i)) = 0 \quad .$$

Locally, there is then a function

$$W(x^1,\ldots,x^n)$$

of x-variables, such that:

$$d\alpha^*(W) = \alpha^*(y_i)d(\alpha^*(x^i)) \quad . \tag{2.6}$$

We can interpret this as follows:

$$\alpha^*\left(y_i - \frac{\partial W}{\partial x^i}\right) = 0 \tag{2.7}$$

Here is the best way of putting this result for the purposes of mechanics:

<u>Theorem 2.2</u>. Let $\alpha: N \to M$ be a Lagrangian submanifold of the symplectic manifold M, and let (y_i, x^i) be a canonical coordinate system for M. Then, there is a real-valued function

$$W(x^1,\ldots,x^n)$$

of the x-variables such that $\alpha(N)$ is defined locally by the relations

$$y_i - \frac{\partial W}{\partial x^i} = 0 \tag{2.8}$$

Conversely, each submanifold defined by a relation of form
(2.8) with W an arbitrary function, is Lagrangian. In parti-
lar, the transversal Lagrangian submanifolds are parameterized
by functions $W(\ )$ of n real variables. Two such functions
W, W' determine the same Lagrangian submanifold if and only if
they differ by a constant.

Proof. We have already seen that a relation of type
(2.8) follows (locally) from a hypothesis that $\alpha(N)$ is a
Lagrangian submanifold. We must show that $\alpha(N)$ is actually
defined by this relation. This hinges on proving that $\alpha(N)$
is defined locally by this relation. Since N is dimension
and α is a submanifold map, this would follow from the
Implicit Function Theorem if it could be proved that the
functions

$$y_i - \frac{\partial W}{\partial x^i}$$

were functionally independent. Again, using the implicit
function theorem, prove this by showing that their differentials are linearly independent. But,

$$d\left(y_i - \frac{\partial W}{\partial x^i}\right) = dy_i - \frac{\partial^2 W}{\partial x^j \partial x^i}\, dx^2 \quad ,$$

which makes it evident that they are linearly independent.
That $\alpha(N)$ is determined locally by W (up to a constant
change in W) is obvious also from (2.8).

<u>Definition</u>. The function $W(x^1,\ldots,x^n)$ which determines the
Lagrangian submanifold $\alpha(N)$ locally by relation (2.8) is
called the <u>generating function</u> of the Lagrangian submanifold.

3.   LAGRANGIAN FIBER SPACES

As a sideline, I will now present a coordinate-free
version of the results of Section 2. Let M continue as a
symplectic manifold, with ω as symplectic form.

<u>Definition</u>. A map

$$\pi: M \rightarrow X$$

between manifolds is said to define a <u>Lagrangian fiber space</u> if
the following conditions are satisfied:

a) π is a submersion map in the sense of differential
   topology, i.e.,

$$\pi_*(T(M)) = T(X)$$

b) Each fiber of π is a Lagrangian submanifold of M.

<u>Remark</u>. Here is one example:

$M = T^d(X)$, the cotangent bundle of X, with
$\pi: T^d(X) \rightarrow X$ the vector-bundle projection map which assigns to
each covector the point of X to which it is attached. One

can prove that each Lagrangian fiber space is <u>locally</u> equivalent
to this one. (This result is basically due to Lie, with his
theory of what he called "function groups".)

Here is a useful way to characterize a Lagrangian fiber
space.

<u>Theorem 3.1</u>. Use the given symplectic structure to define, in
the usual way, a Poisson bracket structure on the functions on
M. Let

$$\pi: M \to X$$

be an arbitrary submersion map, such that:

$$\dim X = \frac{1}{2} \dim M \quad .$$

Then, π is a Lagrangian fiber space if and only if the following condition is satisfied:

$$\{\pi^*(f_1), \pi^*(f_2)\} = 0 \tag{3.1}$$

$$\text{for } f_1, f_2 \varepsilon F(X) \quad .$$

<u>Proof</u>. For $f \varepsilon F(M)$, let A_f be the vector field
such that:

$$df = A_f \lrcorner \omega \tag{3.2}$$

Recall that the Poisson bracket is defined as follows:

$$\{f_1, f_2\} = -A_{f_1}(f_2) \tag{3.3}$$

For $f \in F(X)$, consider

$$A_{\pi^*(f)} \quad .$$

Using (3.2) we have:

$$d\pi^*(df) = A_{\pi^*(f)} \lrcorner \omega \tag{3.4}$$

Now, suppose that π is Lagrangian. This means that each fiber of π is a submanifold of half the dimension of M, on which the form ω is zero. Given $p \in M$, let $V \subset M_p$ be the linear subspace tangent to the fiber of π through p. Then, the "Lagrangian" condition means that

$$\omega(V_p, V_p) = 0 \quad ,$$

i.e., V_p is an <u>isotropic subspace</u> of ω. It also follows from (3.4) that

$$0 = d\pi^*(df)(V_p) = \omega(A_{\pi^*(f)}(p), V_p)$$

Since V_p is a <u>maximal</u> isotropic subspace, we have:

$$A_{\pi^*(f)}(p) \in V_p \quad .$$

Since this holds at each point p of M, $A_{\pi^*(f)}$ is tangent to <u>all</u> the fibers of π. This implies that

$$0 = A_{\pi^*(f)}(\pi^*(f_1)) = \quad , \text{ using } (3.3),$$

$$\{\pi^*(f_1), \pi^*(f)\} \quad ,$$

which proves (3.1).

The proof that, conversely, (3.1) implies the "Lagrangian" nature of π is a straightforward reversal of this argument, and is left to the reader.

<u>Theorem 3.2</u>. Let $\pi: M \to X$ be a Lagrangian fiber space. The
<u>locally</u>, there is a one-form θ such that:

$$d\theta = \omega \tag{3.5}$$

$$\theta = 0 \text{ when restricted to the fibers of } \pi. \tag{3.6}$$

<u>Proof</u>. This readily follows from a theorem well-known
in the 19-th century, and presumably proved first by Lie.
(Perhaps, even by Pfaff.) Let (x^i) be a coordinate system
for X. Then, Theorem 3.1 implies that

$$\{\pi^*(x^i), \pi^*(x^j)\} = 0 , \tag{3.7}$$

where $\{ , \}$ is the Poisson bracket defined by the symplecti
structure. (Lie says that a set of functions satisfying condition (3.7) are <u>in involution</u>.) Lie proves (e.g., see Vol.
of his "Transformationsgruppen" treatise) that there are then
functions y_i on M so that

$$\omega = dy_i \wedge \pi^*(dx^i) . \tag{3.8}$$

θ can obviously be chosen now as follows.

$$\theta = y_i \, d\pi^*(x^i) . \tag{3.9}$$

<u>Exercise</u>. Provide a modern proof of this result.

<u>Definition</u>. A one-form θ satisfying properties (3.5) and (3.6) is called a <u>contact form associated to the Lagrangian fibration</u>.

<u>Theorem 3.3</u>. Let θ be a contact form associated with the Lagrangian fibration of M. Let,

$$\gamma, \gamma_1 : X \to M$$

be cross-section maps of the fiber space such that:

$$\gamma^*(\theta) = \gamma_1^*(\theta) \quad . \tag{3.10}$$

Then (assuming also that X is connected), either

$$\gamma = \gamma_1 \quad .$$

or

$$\gamma(x) \neq \gamma_1(x) \tag{3.11}$$

$$\text{for } \underline{all} \ x \ \varepsilon \ X \quad .$$

$\underline{Proof}$. Let (x^i), $1 \le i,j \le n = \dim X$, be a coordinate system for X. Then, any covector of M which is zero on the fibers of π is a linear combination of the

$$d\pi^*(x^i) \quad .$$

In particular, we have:

$$\theta = y_i \, d\pi^*(x^i) \quad , \tag{3.12}$$

where (y_i) are a set of functions on M.

We then have:

$$\omega = d\theta = dy_i \wedge d\pi^*(x^i) \quad , \qquad (3.13)$$

Since ω is of maximal rank, (3.13) implies that:

$(y_i, \pi^*(x^i))$ forms a coordinate system for M.

Since γ, γ_1 are cross-sections, we have:

$$\pi\gamma = \text{identity} \quad .$$

Hence,

$$\gamma^*\pi^* = \text{identity} \quad . \qquad (3.14)$$

Combine (3.10)-(3.14):

$$\gamma^*(\theta) = \gamma^*(y_i)dx^i = \gamma_1^*(\theta) = \gamma_1^*(y_i)dx^i \quad ,$$

hence,

$$\gamma^*(y_i) = \gamma_1^*(y_i) \quad . \qquad (3.15)$$

Use (3.13) again:

$$\gamma^*(\pi^*(x^i)) = x^i = \gamma_1^*(\pi^*(x^i)) \qquad (3.16)$$

(3.15)-(3.16) imply that γ, γ_1 agree <u>locally</u>. In particular,
the points of X where they agree forms an open subset of X.
It is also obviously either a non-empty closed subset of X,
since γ, γ_1 are continuous, or (3.11) is satisfied. Connec-
tivity of X now implies that

$$\gamma = \gamma_1 \quad .$$

We now ask: Given a one-form θ on X, is there a cross-section

$$\gamma: X \to M$$

such that

$$\gamma^*(\theta) = \alpha \quad ? \tag{3.17}$$

We shall now introduce a global property of the contact form θ which will imply the existence of such a γ. To this end, work with a coordinate system

$$(x^i)$$

for X. As we have seen in the proof of Theorem 3.3, there are then functions y_i on M such that:

$$\theta = y_i \, d\pi^*(x^i) \tag{3.18}$$

Since $d\theta = \omega$ is of maximal rank, the functions $(y_i, \pi^*(x^i))$ are functionally independent. In particular, the:

> y_i restricted to the fibers of π form
> local coordinate systems for the fibers.

Here is the property we need:

$$
\boxed{
\begin{array}{l}
\text{The functions } y_i \text{ form a } \underline{\text{coordinate system}} \\[4pt]
\underline{\text{globally}} \text{ for fibers of } \pi.
\end{array}
}
\tag{3.19}
$$

<u>Exercise</u>. Show that the property (3.19) is independent of the coordinates (x^i) used for the base.

<u>Exercise</u>. Show that a necessary condition that a Lagrangian fibration admit an associated contact structure is that each fiber of π is non-compact. (Of course, if (3.19) is satisfied, each fiber is, in fact, R^n.)

<u>Theorem 3.4</u>. Keep the hypotheses of Theorem 3.3. Suppose, in addition, that (3.19) is satisfied. Then,

$$\gamma_1 = \gamma_2 \ .$$

<u>Exercise</u>. Prove Theorem 3.4.

<u>Theorem 3.5</u>. Suppose that (3.19) is satisfied, and that α is a one-form on X. Then, there is a uniquely defined cross-section map

$$\gamma: X \to M$$

such that

$$\gamma^*(\theta) = \alpha \ . \tag{3.2}$$

<u>Proof</u>. We first prove that γ exists in each coordinate neighborhood. Suppose (x^i) is a coordinate system for X. Let y_i be the functions on M such that:

$$\theta = y_i \, d\pi^*(x^i) \ .$$

The hypothesis (i.e., (3.19)) assumes that

$$(x^i, y_i)$$

form a coordinate system for M.

Suppose that:

$$\alpha = a_i \, dx^i \qquad\qquad (3.21)$$

$\gamma^*(\theta) = \alpha$ is equivalent to the following conditions:

$$\gamma^*(y_i) = a_i \qquad\qquad (3.22)$$

In other words, $\gamma(X)$ can be defined as the subset of
points of M satisfying the following condition:

$$y_i - a_i(x) = 0 \qquad\qquad (3.23)$$

Condition (3.23) can obviously be satisfied, hence
γ exists in each coordinate neighborhood. In the overlap of
two coordinate neighborhoods the cross-sections constructed
separately must agree (by Theorem 3.5), hence, they define a
global cross-section satisfying (3.20).

Remark. We see that, under the hypotheses of Theorem 3.5,
the space of cross-section maps $\gamma: X \to M$ is in one-one correspondence with the space of one-forms α on X. There is, in
fact, a more direct way of seeing this:

Exercise. Under the hypotheses of Theorem 3.5, show that the
fiber space (M, X, π) is equivalent to the cotangent vector
bundle $(T^d(X), X)$.

Theorem 3.6. A cross-section map $\gamma: X \to M$ is a Lagrangian submanifold of the symplectic form ω if and only if the one form α on X associated to it by formula (3.20) is closed. In other words, the Lagrangian cross-section are in one-one correspondence with the closed one-forms.

Proof. "α a closed one-form" means, of course, that $d\alpha = 0$. Now,

$$\alpha = \gamma^*(\theta) ,$$

hence,

$$d\alpha = \gamma^*(d\theta) = \gamma^*(\omega) ,$$

which makes the proof obvious.

This result gives us a way of parameterizing the Lagrangian submanifolds of M which happen to be given direc as cross-section maps of the Lagrangian fibration. How does one tell if a general Lagrangian submanifold arises in this way? Here is one answer:

Definition. A submanifold map

$$\phi: N \to M$$

is transversal to the fiber space map

$$\pi: M \to X$$

if the following condition is satisfied:

For each point p ε N,

$$\phi_*(N_p) \cap V_{\phi(p)} \; = \; 0 \; , \tag{3.24}$$

where $V_{\phi(p)}$ denotes the subspace
of the tangent vectors to M at
$\phi(p)$ which are tangent to the fiber
of π through $\phi(p)$. (See drawing.)

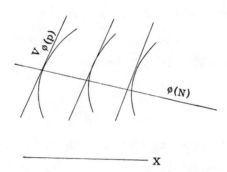

In words, the submanifold $\phi(N)$ cuts
"transversally" across the fibers of π.

In case π is a Lagrangian fiber space, and φ is a
Lagrangian submanifold, both N and the fibers of π have
dimension one-half of the dimension of M. Hence, condition
(3.24) implies that:

$$M_{\phi(p)} \; = \; \phi_*(N_p) \oplus V_{\phi(p)}$$

Hence,

$$\pi\phi: M \to X$$

is a <u>local diffeomorphism</u>. If $\pi\phi$ is a "global" diffeomorphism
then the submanifold ϕ can be <u>reparameterized</u> to be a cross-
section. Hence, modulo certain possible global complications
(which are actually interesting in various contexts!), a <u>trans-</u>
<u>versal Lagrangian submanifold</u> is parameterized by a closed one-
form on X. Locally, then, transversal Lagrangian submanifolds
are parameterized by functions W on X, i.e., by functions
$W(x^1,...,x^n)$ of the "base-like" variables. This identification returns us to the local (and classical) situation discussed in Section 2, with W thought of as the <u>generator</u> of the
Lagrangian submanifold.

After this diversion into the general formalism, let
us turn to the main topic of this chapter, the study of
canonical transformations. From now on in this chapter, I
will actually work only with local canonical coordinates. The
reader will probably see how to translate the ideas into a
global, coordinate-free form, if he should so desire, using the
"Lagrangian fibration" formalism sketched above.

4.   CANONICAL TRANSFORMATIONS AND THEIR GENERATING
        FUNCTIONS

<u>Warning</u>. The notation will now change slightly.

Let us first recall the general definition of a "canonical transformation". Let M and M' be two symplectic

manifolds. Let

$$\omega, \omega'$$

be the closed two-forms on M and M' which define the symplectic structures.

Definition. A diffeomorphism

$$\phi: M \to M'$$

is a canonical transformation if:

$$\phi^*(\omega') = \omega \quad . \tag{4.1}$$

Warning. One frequently finds such a map called a contact transformation. This is very confusing terminology, since it is not the sense in which Lie first defined "contact transformation". Lie's use of the term is based on a very beautiful geometric insight (see my "Geometry, physics and systems", for a brief description), and I regard it as a barbarism to confuse his terminology.

Let us derive conditions that such a map ϕ must satisfy. Let

$$(\text{gr } \phi): M \to M \times M'$$

be the graph of ϕ, i.e.,

$$(\text{gr } \phi)(p) = (p, \phi(p)) \tag{4.2}$$

$$\text{for } p \in M \quad .$$

Put the following symplectic form on $M \times M' \equiv M''$:

$$\omega'' = \omega - \omega' . \tag{4.3}$$

Then, it follows from (7.1) and (7.2) that:

$$(gr \; \phi)^*(\omega'') = 0 \tag{4.4}$$

Condition (4.4) says that $gr(\phi)$ defines a <u>Lagrangian submanifold</u> of the symplectic structure on M''. Thus, the problem of constructing canonical transformations is equivalent to the problem of finding Lagrangian submanifolds of a certain type. As we have seen in previous sections, Lagrangian submanifolds are parameterized by functions of n variables, called the <u>generating functions</u>. We will now see how to construct (in local canonical coordinates) such generating functions for canonical transformations. I will not use the general formalism developed in previous sections (although it is there in the background) since I want to develop the ideas in such a way as to facilitate the calculations of examples planned for later.

Let M'' be a symplectic manifold of dimension $4n$. Let ω'' be its symplectic form. Let

$$(q^i, p_i, x^i, y_i)$$

be a canonical coordinate system for M'', with:

$$\omega'' = dp_i \wedge dq^i - dy_i \wedge dx^i \tag{4.5}$$

Consider a function

$$S(q,y) \tag{4.6}$$

of the indicated variables. Denote by Q, P, X, Y the space of the variables q, p, x, and y, respectively. Thus,

$$M' = Q \times P \times X \times Y \quad . \tag{4.7}$$

Define a map

$$\gamma: Q \times Y \to M''$$

as follows:

$$(q,y) \to \left\{ \begin{array}{l} q \\[1mm] y \\[1mm] p_i = \dfrac{\partial S}{\partial q^i} \\[3mm] x^i = \dfrac{\partial S}{\partial y_i} \end{array} \right\} = (q,y) \tag{4.8}$$

Then,

$$\gamma^*(p_i dq^i + x^i dy_i) = \frac{\partial S}{\partial q^i} dq^i + \frac{\partial S}{\partial y^i} dy^i = dS \quad .$$

Hence,

$$\gamma^*(\omega'') = \gamma^*(d(p_i dq^i + x^i dy_i))$$

$$= d\gamma^*dS = \gamma^*ddS$$

$$= 0 \tag{4.9}$$

Then, γ defines a <u>Lagrangian submanifold</u> of the symplectic manifold M''. It is said to be the <u>Lagrangian submanifold</u> <u>generated by the function</u>

$$S: Q \times Y \to R \quad . \tag{4.10}$$

Here, now, is the main concept:

<u>Definition</u>. A canonical map

$$\phi: Q \times P \to X \times Y$$

is said to be <u>generated</u> by the function S if the submanifo

$$\text{gr} (\phi): Q \times P \to Q \times P \times X \times Y \equiv M''$$

is locally the same submanifold as $\gamma(Q \times Y)$.

<u>Theorem 4.1</u>. Let S be a given function on $Q \times Y$. The necessary and sufficient condition that it generate (locally) a canonical map is that:

$$\det \left(\frac{\partial^2 S}{\partial q^i \partial y_j} \right) \neq 0 \tag{4.11}$$

<u>Proof</u>. Suppose that it does generate the map ϕ, i.e

$$\text{gr} (\phi) - (Q \times P) = \gamma(Q \times Y) \quad ,$$

where γ is defined by (4.8).

Now the functions

$$(p_i, q^i)$$

are functionally independent when restricted to the submanifold

$$gr \ (\phi) \ (Q \times P)$$

Hence, they must also be functionally independent when restricted to $\gamma(Q \times Y)$. By (4.8), the p_i are equal to

$$\frac{\partial S}{\partial q^i}$$

Hence,

$$dq^i, \ d\left(\frac{\partial S}{\partial q^i}\right)$$

must be linearly independent. The condition for this is that
(4.11) hold.

These steps are reversible to prove the converse.

In conclusion, in this section we have covered the
traditional material concerning the "generating function" of
a canonical transformation. There are certain "singular" sorts
of canonical transformations which have been left out. For
the complete story (at least locally), see Whittaker's "Analytical Dynamics", Chapter 11, and p. 153 of GPS. We now proceed on to the material which is of the greatest importance
for quantum mechanics.

5.   CANONICAL TRANSFORMATIONS GENERATED BY COMPLETE
        SOLUTIONS OF THE HAMILTON-JACOBI EQUATION:
        ACTION-ANGLE VARIABLES

Keep the notation of Section 4. Suppose that

$$H(q,p)$$

is a function on $Q \times P = M$, serving as a <u>Hamiltonian functi</u>
of a mechanical system. Let

$$S: Q \times Y \to R$$

be a function of the variables (q,y).

<u>Definition</u>. S is a <u>complete solution of the Hamilton-Jacobi</u>
<u>equation associated with</u> H if the following conditions are
satisfied:

$$H\left(q, \frac{\partial S}{\partial q}(q,y)\right) \quad \text{is a function of y \underline{alone}} \quad (5.1)$$

$$\det\left(\frac{\partial^2 S}{\partial q^i \partial y_j}\right) \neq 0 \qquad (5.2)$$

Suppose that S is such a complete solution. As we
have seen in Section 4, S generates a canonical transformation

$$\phi: Q \times P \to X \times Y$$

and a map

$$\gamma: Q \times Y \to M'' \quad ,$$

defined by formulas (4.8), such that

$$\text{gr } (\phi)(Q \times P) \ = \ \gamma(Q \times Y) \quad .$$

Set:

$$h(y) \ = \ H\left(q, \frac{\partial S}{\partial q} (q,y)\right) \tag{5.3}$$

Consider h as a function on $X \times Y$. (In this case, of course, h is a <u>function of</u> y <u>alone</u>.)

<u>Theorem 5.1</u>.

$$\phi^*(h) \ = \ H \tag{5.4}$$

<u>Proof</u>. Define

$$h'': M'' \to R$$

by the following formula.

$$h''(q,p,x,y) \ = \ -H(q,p) - h(y) \quad .$$

Now, use formulas (4.8) which define γ:

$$\gamma^*(h'') \ = \ H\left(q, \frac{\partial S}{\partial q}\right) - h(y)$$

$$= \ 0, \quad \text{by } (5.3) \tag{5.5}$$

We can reinterpret (5.5) by saying that the function h'' is zero on the submanifold

$$\alpha(Q \times Y) \quad .$$

But, $\alpha(Q \times Y) = gr(\phi)(Q \times P)$,   hence

      h'' is zero on the graph of ϕ.

This means that (5.4) is satisfied.

    Here is the most important consequence of Theorem 5.1. Consider the solutions of the Hamilton equations, with Hamiltonian H:

$$\frac{dq}{dt} = \frac{\partial H}{\partial p}$$

$$\frac{dp}{dt} = \frac{\partial H}{\partial q}$$

$$\text{(5.6)}$$

We know that the condition that a curve

$$t \to (q(t),\, p(t))$$

is a solution of (5.6) is that the curve

$$t \to (q(t),\, p(t),\, t)$$

is a characteristic curve of the two-form

$$dp_i \wedge dq^i - dH \wedge dt \quad .$$

    Consider

$$\phi(q(t),\, p(t)) = (x(t),\, y(t)) \qquad \text{(5.7)}$$

Because of (5.4) and the canonical map condition

$$\phi^*(dy_i \wedge dx^i) = dp_i \wedge dq^i \quad ,$$

we see that

$$t \rightarrow (x(t), y(t), t) \qquad\qquad (5.8)$$

is a characteristic curve of

$$dy_i \wedge dx^i - hdt \quad ,$$

i.e., the curve (5.8) is a solution of:

$$\frac{dx}{dt} = \frac{\partial h}{\partial y} \quad ; \qquad \frac{dy}{dt} = -\frac{\partial h}{\partial x} \quad . \qquad\qquad (5.9)$$

(Canonical transformations preserve the "Hamiltonian" nature of differential equations.)

Now, h is a function of y alone. Hence, Equations (5.9) take the following form:

$$\frac{dy}{dt} = 0 \qquad .$$

$$\frac{dx}{dt} = \frac{\partial h}{\partial y} (y) \quad ,$$

i.e.,

$$y(t) = y_0$$

$$\qquad\qquad (5.10)$$

$$x(t) = \frac{\partial h}{\partial y} (y_0)t + x(t_0)$$

One can also regard the (x,y) more traditionally as "new coordinates" for the phase space of variables

$$(p,q) \quad .$$

These coordinates then have the following property:

The y_i are constants of motion of the
Hamilton equations (5.6). The x^i
increase linearly.

Such systems of coordinates are called <u>action-angle</u>
<u>variables</u> in the physics literature. (See Goldstein [1].)
Finding them is equivalent to "solving" the problem.

6.   QUANTIZATION VIA A COMPLETE SOLUTION OF THE HAMILTON-
      JACOBI EQUATION

We shall now attempt a "quantum" interpretation of
the geometric ideas sketched in the previous two sections.
Start with a configuration space manifold Q. $T^d(Q)$, its
<u>cotangent bundle</u>, is the <u>phase space</u> of the mechanical system.
      Suppose given a set of classical observables $\underline{O}$,
given as functions on $T^d(Q)$. Among these, single out one,
denoted by H, called the <u>total energy</u> of the system. H
determines the dynamics of the system, via Hamilton's equations
and the symplectic structure on $T^d(Q)$.
      Suppose given a <u>quantization</u> as a mapping

      $\underline{O} \rightarrow$ (differential operators on Q)   .

Usually, the most important practical problem is to find the
<u>spectrum</u> of the operator corresponding to H.

Often, one can regard the "classical" mechanics problem as "solved" when a canonical transformation of the type discussed in Sections 4 and 5 (i.e., generated by a complete solution of the Hamilton-Jacobi equation) can be found. We would like to use this method to deal with the quantum problem also.

Here is one possible way to do this. Suppose X, Y are manifolds, such that

$$X \times Y$$

is a symplectic manifold. Suppose coordinates

$$(x^i) \quad \text{for} \quad X ,$$

$$(y_i) \quad \text{for} \quad Y ,$$

are given, such that the symplectic form for $X \times Y$ is:

$$dy_i \wedge dx^i .$$

Suppose given a canonical map

$$\phi : T^d(Q) \to X \times Y$$

such that:

$$\phi^*(h) = H , \tag{6.1}$$

where

$$h \quad \text{is a function of} \quad y \quad \text{alone.} \tag{6.2}$$

We can then consider the observables $\underline{O}$ as identified with the functions

$$\phi^{-1}*(\underline{O})$$

on $X \times Y$. In favorable circumstances, they can be quantized
(e.g., using the Schrödinger rules) as differential operators
on X.

In particular, consider the Hamiltonian, or total
energy, H. It corresponds, because of (6.1), to h.

Suppose that h, as a function of y, is a polynomial
in y. Then, following the Schrödinger rules, h is assigned
a skew-Hamiltonian differential operator on X. In the variables (x^i), this differential operator has constant coeffici
ents, hence its spectrum can be readily found. Hopefully,
this will be identical with the spectrum of the differential
operator on Q corresponding to its H.

Remark. I say "hopefully", because it is simply not known
whether a canonical transformation of the classical phase
space "corresponds" to a unitary transformation of the Hilbert
space of quantum states and observables. From my point of vie
(described in VB, Vol. 2), this is really the fundamental
(and unsolved) "problem" concerning the relations between
classical and quantum mechanics. However, for practical purposes, it is presumably reasonable to proceed as if this were
true.

A procedure analogous to this one for the quantization
of classical field theories has been suggested by Faddeev
and Zakharov. They have worked out the details of

the construction of ϕ (via an "inverse scattering problem", rather than solving the H-J equation) in two cases: $T^d(Q)$ = space of solutions (satisfying appropriate smoothness and asymptotic properties) of the Korteweg-DeVries equations and the sine-Gordon equation. In fact, it is their work (which I learned about in lectures by Faddeev at Harvard University in June, 1975) which has been part of what motivated me to write down in detail this exposition of the Hamilton-Jacobi method!

7.   THE HARMONIC OSCILLATOR

In the next sections I will work out examples of canonical transformations generated by complete solutions of the Hamilton-Jacobi equation. The system the physicists call the harmonic oscillator is probably the simplest non-trivial example. (The "harmonic oscillator" is the same as the "spring", "linear pendulum", etc., encountered in elementory mechanics books.)

It has a one-dimensional configuration space Q, diffeo-morphic to the real numbers R. Denote the variable in Q by q. P, X and Y (following the notation of Section 5) are also taken as one-dimensional. The Hamiltonian H takes the following form:

$$H = p^2 + q^2 \qquad (7.1)$$

The Hamilton-Jacobi equation corresponding to (7.1) is:

$$\left(\frac{\partial S}{\partial q}\right)^2 + q^2 = h(y) \qquad (7.2)$$

This is to be solved for a function

$$S(q,y) \quad ,$$

where $h(y)$ is a function to be found later.

Rewrite (7.2) as

$$\frac{\partial S}{\partial q} = \sqrt{h - q^2} \quad ,$$

or

$$S(q,y) = \int (h(y) - q^2)^{1/2} \, dq \quad . \qquad (7.3)$$

Following the general development of Section 5, we next find a map

$$\gamma: Q \times Y \rightarrow Q \times P \times X \times Y \equiv M'' \quad ,$$

via the following formula (a specialization of (5.8):

$$\gamma(q,y) = \left(q, \ y, \ p = \frac{\partial S}{\partial q}, \ x = \frac{\partial S}{\partial y}\right) \quad . \qquad (7.4)$$

Now, let us use (7.3) to calculate $P(q,y)$ and $x(q,y)$

$$P = \frac{\partial S}{\partial q} = \sqrt{h(y) - q^2} \qquad (7.5)$$

$$x = \frac{\partial S}{\partial y} = \int (h(y)-q^2)^{-1/2} \frac{1}{2} \frac{dh}{dy} dq \qquad (7.6)$$

Make, as usual, the substitution:

$$q = \sqrt{h} \sin (z) \quad , \qquad (7.7)$$

$$dq = \sqrt{h} \cos (z) dz \qquad (7.8)$$

Combine $(7.6)-(7.8)$:

$$x = \frac{1}{2} \frac{dh}{dy} \int h^{-1/2}((\cos z)^{-1}) \sqrt{h} \cos (z) dz$$

$$= \frac{1}{2} (z-k(y)) \frac{dh}{dy} \qquad (7.9)$$

$(k(y)$, for the moment, is another arbitrary function of y.)
We can now substitute (7.7) into (7.9):

$$x = \frac{1}{2} \left(\sin^{-1} \frac{q}{\sqrt{h}} - k(y) \right) \frac{dh}{dy} \qquad (7.10)$$

In order to make what we have done as clear as possible,
let us summarize as follows:

Theorem 7.1. Let $M'' = Q \times P \times X \times Y$. The complete solution
(7.3) of the harmonic oscillator Hamilton-Jacobi equation determines a Lagrangian submanifold map

$$\gamma: Q \times Y \to M''$$

as follows:

$$\gamma(q,y) = \left(q, p = \sqrt{h(y)-q^2} \ , \ x = \frac{1}{2}\left(\sin^{-1}\frac{q}{\sqrt{h}} - k(y)\right)\frac{dh}{dy} \ , \ y\right)$$

$$(7.11)$$

We can use formula (7.11) to compute the canonical transformation

$$\phi: Q \times P \to X \times Y$$

generated by S. To do this, let us solve the relation

$$p = \sqrt{h(y)-q^2} \quad ,$$

namely,

$$h(y) = p^2 + q^2 \tag{7.12}$$

Combine (7.10) and (7.12)

$$x = \frac{1}{2}\left(\sin^{-1}\left(\frac{q}{\sqrt{p^2+q^2}}\right) - k(y)\right)\frac{dh}{dy} \tag{7.13}$$

The following choice of $h(y)$ is the obvious one required to write these formulas in an interesting global way:

$$k(y) = y^2 \tag{7.14}$$

With this choice, we have:

$$x = \frac{1}{2}(\sin^{-1}(q/y) - k(y))2y \quad ,$$

or

$$\sin^{-1}\left(\frac{q}{y}\right) \;=\; \frac{x}{y} + k(y) \quad,$$

or

$$q \;=\; y \sin\left(\frac{x}{y} + k(y)\right) \qquad\qquad (7.15)$$

$$p^2 \;=\; y^2 - q^2$$

$$ \;=\; y^2 - y^2 \sin^2 (x+k(y))$$

$$ \;=\; y^2 \cos^2 (x+k(y)) \quad,$$

or

$$p \;=\; y \cos (x+k(y)) \qquad\qquad (7.16)$$

Let us sum up as follows:

<u>Theorem 7.2.</u> Let X be the real numbers, parameterized by
x. Let

$$X \times Y$$

be its cotangent bundle, <u>minus the zero cross-section</u>, parameterized by (x,y), $x \in R$, $y \in R-(0)$. Let $k(y)$ be an arbitrary
function of y. Let Q be another copy of R, with $Q \times P$
its cotangent bundle. Let

$$h(y) \;=\; y^2 \qquad\qquad\qquad\qquad (7.17)$$

$$H(p,q) \;=\; p^2 + q^2 \qquad\qquad\qquad (7.18)$$

be Hamiltonian functions of mechanical systems, with X and Q, respectively, configuration spaces. Define a map

$$\phi^{-1}: X \times Y \to Q \times P \qquad (7.19)$$

as follows:

$$\phi^{-1}(x,y) \;=\; \left(y \,\sin\!\left(\frac{x}{y} + k(y)\right),\; y\,\cos\!\left(\frac{x}{y} + k(y)\right) \right) \qquad (7.20)$$

Then, ϕ^{-1} is a canonical map between the symplectic structure i.e.,

$$(\phi^{-1})^*(dy \wedge dx) \;=\; dp \wedge dq \qquad (7.21)$$

Also,

$$(\phi^{-1})^*(H) \;=\; h \qquad (7.22)$$

In particular, ϕ^{-1} sends solutions of the Hamilton equations with Hamiltonian h into solutions of the Hamilton equations with Hamiltonian   .

Exercise. Prove these results independently of the general theory by verifying formulas (7.20) and (7.21) directly, using (7.19).

Remark. h is the Hamiltonian of a free particle. The energy spectrum of its quantum-mechanical energy operator is continuous. I will attempt to explain the quantum mechanical nature of these formulas--in particular, why the many-one nature of the mapping ϕ^{-1} forces a discrete energy spectrum on H-- at a later point.

This way of looking at the solution of the problem also suggests various generalizations. Here is one:

8.   THE HAMILTON-JACOBI EQUATION FOR THE PENDULUM

Keep Q, P, X, Y as in Section 7. Generalize as follow

$$H(q,p) = \frac{1}{2} p^2 + V(q) \tag{8.1}$$

The Hamilton-Jacobi equation is now:

$$\left(\frac{\partial S}{\partial q}\right)^2 + V(q) = h(y) \quad ,$$

or

$$p = \frac{\partial S}{\partial q} = \sqrt{h(y) - V(q)} \quad , \tag{8.2}$$

$$S = \int \sqrt{h(y) - V(q)} \, dq \quad . \tag{8.3}$$

Hence,

$$x = \frac{\partial S}{\partial y} = \int \frac{1}{2} (h(y) - V(q))^{-1/2} \frac{dh}{dy} \, dq \tag{8.4}$$

Also,

$$p^2 + V(q) = h(y) \quad . \tag{8.5}$$

Formulas (8.2)-(8.5) define the canonical transformation ϕ locally. (In classical terminology, one could say that ϕ is defined _implicitly_.) We would like to know when formulas analogous to (7.14) and (7.15) can be found, so that ϕ^{-1} can be found _globally_. I do not know of any work on this general problem. One can look through a table of Elliptic Integrals (e.g., Byrd and Friedman [1]) and see that for many potential functions $V(q)$, this can be done in terms of Elliptic Functions. For example, consider the pendulum:

$$V(q) = -\cos q . \tag{8.6}$$

Combine (8.4) and (8.6):

$$x = \frac{1}{2} \frac{dh}{dy} \left(\int_0^q (h(y) + \cos q)^{-1/2} dq + j(y) \right)$$

= , using formula 289 on page 174 of Byrd and
    Friedman [1],

$$\frac{1}{2} \frac{dh}{dy} (g \ sn^{-1} (\sin \beta, k) + k(y)) , \tag{8.7}$$

with:

$$g = \frac{2}{\sqrt{h(y)+1}} \tag{8.8}$$

$$k^2 = \frac{2}{h(y)+1}$$

$$\beta = q/2 \tag{8.9}$$

$j(y)$ is an arbitrary function of y. $sn (,)$ is the usual Jacobian elliptic function, and $sn^{-1} (,)$ is the inverse function.

Let us now rewrite (8.7) by inverting sn^{-1}:

$$g \ sn^{-1}(\sin \beta, k) = \frac{2x}{(dh/dy)} - j(y)$$

$$\sin \frac{q}{2} = sn \left(g \left(\frac{2x}{(dh/dy)} - j(y) \right), k \right) \qquad (8.10)$$

Now,

$$\sin \frac{q}{2} = \sqrt{\frac{1 - \cos q}{2}}$$

$$= \sqrt{\frac{1 + V(q)}{2}} \quad ,$$

$$V(q) = 2 \sin^2 \frac{q}{2} - 1$$

$$= 2 \left(\sin \left(g \left(\frac{2x}{(dh/dy)} - j(y) \right), k \right)^2 - 1 \right)$$

$$= h(y) - p^2 \quad .$$

These formulas give a reasonably explicit way to define the canonical transformation

$$\phi^{-1} : (x,y) \rightarrow (q,p)$$

in terms of elliptic functions. Unfortunately, they are not
as symmetric as in the linear oscillator case. I plan later
to deal systematically with these problems.

9.   CENTRAL MOTION IN THE PLANE

        Let the configuration space variables be the usual polar
coordinates

        (r,θ) .

p_r, p_θ denote the dual momentum variables. The Hamiltonian
is given by the following formula:

$$H = \frac{1}{2}\left(p_r^2 + \frac{p_\theta^2}{r^2}\right) + V(r) \quad . \tag{9.1}$$

        It is well known that the Hamilton-Jacobi equation associ-
ated with this Lagrangian is solvable by the method of "separation of variables". This means that we look for a complete
solution of the following form:

$$y = (y_r, y_\theta)$$
$$x = (x_r, x_\theta) \tag{9.2}$$

Warning. x and y are not Cartesian coordinates in the
plane. They are related to what are usually called "Keplerian
elements" in the celestial mechanics books.

S, the complete solution of the H-J equation, is of the following form:

$$S(r,\theta,y) = S_1(r,y_r,r_\theta) + S_2(\theta,y_r,y_\theta) \qquad (9.3)$$

Hence, the H-J equation takes the following form:

$$\frac{1}{2}\left(\left(\frac{\partial S_1}{\partial r}\right)^2 + \left(\frac{\partial S_2}{\partial \theta}\right)^2 \frac{1}{r^2}\right) + V(r) = h(y) \qquad (9.4)$$

Multiply by $2r^2$:

$$r^2\left(\frac{\partial S_1}{\partial r}\right)^2 + \left(\frac{\partial S_2}{\partial \theta}\right)^2 + 2r^2 V(r) = 2r^2 h(y) \qquad (9.5)$$

In order that Equation (9.5) be compatible, we must obviously have:

$$S_2 = h_1(y_r,y_\theta)\theta + h_2(y_r,y_\theta) \quad , \qquad (9.6)$$

for some choice h_1, h_2 of functions of the indicated variables.

Substitute (9.6) back into (9.5):

$$r^2 \frac{\partial S_1}{\partial r}^2 h_1^2 + 2r^2 V(r) = 2r^2 h(y) \qquad (9.7)$$

Physically, $h(y)$ represents the <u>energy</u>, h_1 represents <u>angular momentum</u>. (Both are conserved, because the potential is rotationally invariant.) Let us label them as:

$$E, L \quad ,$$

and suppose that they determine a coordinate system for y-spac

We can now rewrite (9.6) and (9.7) as follows, using the new

variables:

$$S_2 = L\theta \tag{9.8}$$

(Set $h_2 = 0$.)

$$r^2 \left(\frac{\partial S_1}{\partial r}\right)^2 + L^2 + 2r^2 V = 2r^2 E \tag{9.9}$$

We now describe Jacobi's canonical transformation,

$$\phi: (r, \theta, p_r, p_\theta) \rightarrow (x_E, x_L, E, L) \quad ,$$

by writing the equations which determine its graph:

$$p = \frac{\partial S}{\partial \theta} = L \tag{9.1}$$

$$p_r = \frac{\partial S}{\partial r} = (2E - 2V - L^2(r^2))^{1/2} \tag{9.1}$$

$$x_E = \frac{\partial S}{\partial E} = \frac{\partial}{\partial E} \int (2E - 2V(r) - L^2/r^2)^{1/2} \, dr$$

$$= \int (2E - 2V(r) - L^2(r^2))^{-1/2} \, dr \tag{9.1}$$

$$x_L = \frac{\partial S}{\partial L} = \theta - L \int (2E - 2V(r) - L^2/r^2)^{-1/2} \, r^{-2} \, dr$$

$$\tag{9.1}$$

Further progress in writing down ϕ explicitly involves doing the integrals on the right hand side of (9.13). There is an interesting qualitative discussion about the possibilities here in Goldstein's "Classical Mechanics", Chapter 3, Section 5. A limited number of potential functions $V(r)$ lead to integrals that can be done by means of trigonometric functions (among these are the two most important examples, the harmonic oscillator, $V = r^2$, and the Kepler-Newton case, $V = $ constant $\times 1/r$), and a few more possibilities are do-able in terms of elliptic functions.

The standard approach to these integrals is to change variables as follows:

$$s = 1/r . \tag{9.14}$$

Remark. That this is a successful choice can be seen on the basis of general principles. The conformal transformation

$$(\theta, r) \rightarrow (\theta, 1/r)$$

of configuration space transforms the Kepler problem into the linear harmonic oscillator. (See IM, Vol. 10.) This is the transformation which "regularizes" the collision in the two-body celestial mechanics problem.

With the substitution (9.14), we have

$$dr = -\frac{1}{s^2} ds ,$$

$$x_E = \int -(2E - 2V(1/s) - L^2 s^2)^{-1/2} s^{-2} \, ds \qquad (9.15)$$

$$x_L = \theta - L \int (2E - 2V(1/s) - L^2 s^2)^{-1/2} \, ds \qquad (9.16)$$

For example, let us do the Kepler case:

$$V(r) = \lambda/s \qquad (9.17)$$

where λ is a constant. Then,

$$x_E = -\int \frac{ds}{s^2 (2E - 2\lambda s - L^2 s^2)^{1/2}} \qquad (9.18)$$

$$x_L = \theta - L \int \frac{ds}{(2E - 2\lambda s - L^2 s^2)^{1/2}} \qquad (9.19)$$

To evaluate these integrals, use the following integration formulas: ("x" denotes dummy integration variables, as usual in calculus.)

$$\int (ax^2 + bx + c)^{-1/2} \, dx = \frac{-1}{(-a)^{1/2}} \sin^{-1}\left(\frac{2ax + b}{(b^2 - 4ac)^{1/2}} \right)$$

$$(9.20)$$

for $a < 0$, $b^2 > 4ac$

$$\int \frac{dx}{x^2 (ax^2 + bx + c)} = -\frac{(ax^2 + bx + c)^{1/2}}{cx} - \frac{b}{2c} \int \frac{dx}{x(ax^2 + bx + c)^{1/2}}$$

$$= -\frac{(ax^2+bx+c)^{1/2}}{cx} - \frac{b}{2c(-c)^{1/2}} \sin^{-1} \frac{bx+2c}{x(b^2-4ac)^{1/2}}$$

(9.21)

To evaluate (9.17) and (9.18) in terms of (9.19), (9.20), make the following substitution:

$$x = s; \quad a = -L^2; \quad b = -2\lambda; \quad c = 2E .$$

$$x_L = \theta + \frac{L}{L} \sin^{-1}\left(\frac{-2L^2s-2\lambda}{(4\lambda^2+8L^2E)^{1/2}}\right)$$

or

$$\frac{2L^2s+2}{(4\lambda^2+8L^2E)^{1/2}} = -\sin(x_L-\theta) ,$$

or, substituting back $s = 1/r$,

$$\frac{2L^2r^{-1}+2\lambda}{(4\lambda^2+8L^2E)^{1/2}} = -\sin(x_L-\theta)$$

(9.22)

$$x_E = \frac{(2sE-2\lambda s-L^2s^2)^{1/2}}{2Es} + \frac{(-2\lambda)}{4E(-2E)^{1/2}} \sin^{-1} \frac{-2\lambda s+4E}{s(4\lambda^2+8EL^2)^{1/2}}$$

or

$$\frac{-2\lambda s+4E}{s(4\lambda^2+8EL^2)^{1/2}} = \sin\left(\frac{4E(-2E)^{1/2}}{2} \frac{(2sE-2\lambda s-L^2s^2)^{1/2}}{2Es} - x_E\right)$$

= , using (9.11) ,

$$\sin\left(\frac{4E(-2E)^{1/2}}{2\lambda} \left(\frac{rp_r}{2E} - s_E \right) \right) \tag{9.23}$$

Unfortunately, these formulas cannot be solved to write ϕ^{-1} by an explicit formula, as we could for the harmonic oscillator. When appropriate, geometric variables (instead of "physical" variables E,L) are chosen (i.e., those parameteriz-ing the ellipses in which the orbits move), these equations can be written in simpler-looking form, and become more-or-less identical with Kepler's laws of planetary motion. (See Gold-stein [1] and Sommerfeld's "Mechanics", Section 46, for details

Chapter 12

INVARIANT INTEGRALS
AND CONTINUUM MECHANICS

1.   INTRODUCTION

The differential equations of motion of a mechanical
system of course have a special structure. It is this structure
that distinguishes "analytical mechanics" from the general
discipline of "non-linear ordinary differential equations".

Traditionally, these special features are related to the
calculus of variations and Hamilton-Jacobi theory. Thus, any
reasonably complete book on mechanics contains extensive
discussion of these topics.

Poincaré and E. Cartan rather emphasized the role that
differential forms played in the study of the equations of
mechanics. They called this topic the theory of "invariant
integrals." Poincaré's work in his "Mecanique Céleste" is rather
casual, but Cartan, in "Lécons sur les invariants intégraux"
provided a magnificent (although eccentric) systematic presen-
tation, which laid the foundation for the present day study
of mechanics in a differential geometric context.

In this chapter, I will go over some of the general
material concerning "invariant integrals", and then consider
some specific applications.

2.   EULERIAN AND LAGRANGIAN INFINITESIMAL GERERATORS OF
     FLOWS

Let X be a manifold. Let t be a real parameter,
varying, say, over

$$-\infty < t < \infty$$

A <u>flow</u> on X is a one-parameter family

$$t \to \phi_t$$

of diffeomorphisms on X, such that the map

$$(t,x) \to \phi_t(x)$$

of

$$R \times X \to X$$

is C^∞. A curve

$$t \to x(t)$$

in X is an <u>orbit</u> of the flow if there is an $x_0 \in X$ such
that

$$x(t) = g(t)(x_0) \tag{2.1}$$

for all t.

A system

$$\frac{dx}{dt} = a(x,t) ; \quad x \in X \tag{2.2}$$

of ordinary differential equations defines such a flow;   the orbits are the solutions of the differential equation.

One can attach to the flow in a natural way two one-parameter families

a) $t \to A_t$

b) $t \to B_t$

of vector fields on   X,   called the Eulerian and Lagrangian infinitesimal generators.   (The terminology arises from fluid mechanics.)   Here are their geometric definitions:

a)   For $x \in X$, $t \in R$, $A_t(x)$ is the tangent vector to the orbit which at time   t   is at the point   x.

b)   For $x \in X$, $t \in R$, $B_t(x)$ is the tangent vector to the curve

$$s \to \phi_t^{-1}(\phi_s(x)) \qquad \text{at} \quad s = t \quad .$$

Let us work out the formulas of Lie derivative by A_t and B_t.   Let $s \to x(s)$ be the orbit of the flow which at time   t   is at   x.   Then

$$x(s) \;=\; \phi_s(x_0)$$

for some $x_0 \in X$.

Also,

$$x(t) \;=\; x \;=\; \phi_t(x_0)$$

Hence,

$$x(s) \;=\; \phi_s(\phi_s^{-1}(x)) \quad .$$

This proves that:

> $A_t(x)$ is the tangent vector to the curve
> $s \to \phi_s(\phi_t^{-1}(x))$ at $s = t$.

In particular, for $f \in F(X)$,

$$A_t(f)(x) \;=\; A_t(x)(f) \;=\; \frac{\partial}{\partial s} f(\phi_s \phi_t^{-1}(x)) \Big|_{s=t}$$

$$=\; \frac{\partial}{\partial s} \phi_t^{-1*}(\phi_s^*(f))(x) \Big|_{s=t}$$

or

> $$A_t(f) \;=\; \phi_t^{-1*} \frac{\partial}{\partial t} \phi_t^*(f)$$

(2.3)

Similarly,

$$B_t(f)(x) \;=\; B_t(x)(f) \;=\; \frac{\partial}{\partial s} f(\phi_t^{-1}\phi_s(x)) \Big|_{s=t}$$

$$=\; \frac{\partial}{\partial s} \phi_s^* \phi_t^{-1*}(f)(x) \Big|_{s=t}$$

or

$$B_t(f) = \frac{\partial}{\partial s} \phi_s^* \phi_t^{-1*}(f) \Big|_{s=t} \qquad (2.4)$$

Note that (2.3) can be written as

$$\frac{\partial}{\partial t} \phi_t^*(f) = \phi_t^*(A_t(f)) \qquad (2.5)$$

(2.4) can be written as:

$$\frac{\partial}{\partial t} (\phi_t^*(f)) = B_t(\phi_t^*(f)) \qquad (2.6)$$

We can use (2.6) to write the ordinary differential
equations (2.2) whose solutions determine the flow.

Suppose that

$$t \rightarrow \phi_t(x_0)$$

is an orbit of the flow. Let

$$(x^i)$$

be a coordinate system for X. Set:

$$x^i(t) = x^i(\phi_t(x_0)) = \phi_t^*(x^i)(x_0) \quad .$$

Let $a^i(x,t)$ be the functions such that

$$A_t = a^i \frac{\partial}{\partial x^i} \qquad (2.7$$

Then,

$$\frac{d}{dt} x^i(t) = \frac{\partial}{\partial t} \phi_t^*(x^i)(x_0)$$

$$= \quad , \text{ using } (2.5),$$

$$\phi_t^*(A_t(x^i))(x_0)$$

$$= \phi_t^*(a^i)(x_0)$$

$$= a^i(x(t),t)$$

This shows that the orbits satisfy the ordinary differential equations (2.2). In other words,

> The Eulerian infinitesimal generator can be read off directly from the differential equation.

Exercise. If ω is an arbitrary differential form on X, show that (2.2) holds with ω replacing f, i.e., that:

$$\frac{\partial}{\partial t} \; \phi_t^*(\omega) \;\; = \;\; \phi_t^*(A_t(\omega)) \tag{2.8}$$

The Eulerian and Lagrangian infinitesimal generators are related via the following relation:

$$\phi_t^{-1*} B_t \phi_t^* \;\; = \;\; \phi_t^{-1*} \frac{\partial}{\partial s} \; \phi_s^* \Big|_{s=t}$$

$$\tag{2.9}$$

$$= \;\; A_t \quad .$$

This relation can also be written as:

$$A_t \;\; = \;\; (\phi_t)_*(B_t) \tag{2.10}$$

3.   INVARIANT INTEGRALS OF FLOWS

Let

$$t \to \phi_t$$

be a flow of the type we have considered in Section 2. In the applications (e.g., to mechanics) the flow will be generated (in the "Eulerian" way) by a differential equation of type (2.2).

Let σ be a submanifold of dimension m of X. Let

$$\sigma_t \;\; = \;\; \phi_t(\sigma)$$

be the transform of σ be the flow. Given a differential m-form ω on X, consider the function

$$t \to \int_{\sigma_t} \omega \qquad \qquad (3.1$$

Definition. The differential form ω is said to be an <u>invar</u> <u>ant integral of the flow</u> if the function (3.1) is actually independent of t, for each m-dimensional submanifold σ of X.

Here is the picture that goes along with this, say, in the case m = 1, dim X = 2

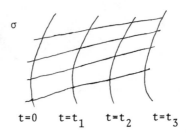

The vertical curve on the left is the initial submanifold. T horizontal curves are the "streamlines" of the flow, i.e., t "orbits", i.e., the solutions of the differential equation (2.1). The other vertical curves are the transforms of σ under the flow at succeeding times

$$t = t_1, t_2, t_3, \ldots$$

Theorem 3.1. Let

$$t \rightarrow A_t \; \varepsilon \; V(X)$$

be the (Eulerian) infinitesimal generator of the flow, as
defined by (2.3). Then, ω is an invariant integral if and
only if

$$A_t(\omega) \;=\; 0 \tag{3.2}$$

for all t .

Proof. We have

$$\int_{\sigma_t} \omega \;=\; \int_{\phi_t(\sigma)} \omega$$

$$=\; \text{, using the transformation law for the}$$
integrals of differential forms,

$$\int_{\sigma} \phi_t^*(\omega) \quad .$$

Hence,

$$\frac{d}{dt} \int_{\sigma_t} \omega \;=\; \frac{d}{dt} \int_{\sigma} \phi_t^*(\omega)$$

$$=\; \int_{\sigma} \frac{\partial}{\partial t} \phi_t^*(\omega)$$

$$= \quad , \text{ using } (2.5) \quad ,$$

$$\int_\sigma \phi_t^*(A_t(\omega))$$

This can be zero if and only if

$$\int_\sigma A_t(\omega) \quad = \quad 0 \tag{3.3}$$

for <u>all</u> submanifolds σ. (3.3) can be satisfied only if (3.2) is satisfied.

Example. <u>Flows which preserve volumes</u>.

Suppose that ω is a <u>volume element</u> differential form on X, i.e., the degree of ω is equal to the dimension of X, and ω is non-zero at each point of X. The condition that ω be an integral invariant is that each diffeomorphism ϕ_t of the flow <u>preserves</u> the <u>volume</u>. For example, if (x^i) is a coordinate system for X, with

$$\omega \quad = \quad dx^1 \wedge \ldots \wedge dx^n \quad ,$$

and if

$$A_t \quad = \quad A_t^i \frac{\partial}{\partial x^i} \quad ,$$

then condition (3.2) amounts to

$$\frac{\partial}{\partial x^i} (A_t^i) \quad = \quad 0 \quad .$$

This is the familiar condition that the <u>divergence of the vector field be zero</u>.

<u>Exercise</u>. Show that ω is an invariant integral of the flow $t \to \phi_t$ if and only if <u>either</u> of the following conditions are satisfied:

$$\phi_t^*(\omega) = \omega$$

for all t .

$$B_t(\omega) = 0 \quad ,$$

where $t \to B_t \in V(X)$ is the Lagrangian infinitesimal generator of the flow.

<u>Warning about terminology</u>. If there were no historical background, a better term than "invariant integral" would be "invariant <u>differential forms</u>". However, Cartan uses this term for a completely different concept, which I will explain later on.

4.   RELATIVE INVARIANT INTEGRALS

Poincaré and Cartan discussed two sorts of invariant integrals called <u>absolute</u> and <u>relative</u>. The former type has been considered in Section 3.

<u>Definition</u>. Let ω be a differential form of degree m on a manifold X, and let $t \to \phi_t$ be a flow on X. Then, ω is called a <u>relative differential invariant of the flow</u> if the following condition is satisfied:

$$\int_{\phi_t(\sigma)} \omega = \int_{\sigma} \omega \qquad (4.1)$$

for each m-dimensional submanifold which is <u>the boundary of a</u> $(m+1)$-<u>dimensional submanifold</u>.

Here is the picture which goes with this.

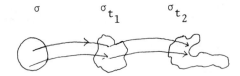

Theorem 4.1. ω is a relative invariant integral if and only if $d\omega$ is an absolute integral invariant.

The <u>proof</u> is a consequence of Stokes' formula, i.e.,

$$\int_{\left(\begin{smallmatrix} \text{boundary of a} \\ \text{submanifold} \end{smallmatrix}\right)} \omega = \int_{(\text{submanifold})} d\omega$$

Example. The traditional example (in the work of Poincaré and Cartan) is that associated with Hamiltonian mechanics. Suppose

$$X = R^{2n} .$$

Let coordinates on X be

$$(p_i, q^i) , \qquad\qquad 1 \le i,j \le n .$$

Let $H(p,q,t)$ be a function of the

(momentum × position × time)

variables. Let $t \to \phi_t$ be the flow generated by the Hamilton equations, with H as Hamiltonian.

Set:

$$\theta = p_i dq^i$$

$$\omega = dp_i \wedge dq^i .$$

Then, θ is a relative, ω and absolute, invariant integral. In this case, since ω defines a symplectic structure on X, another possible terminology would be to say that $t \to \phi_t$ is a symplectic flow.

5.   THE AUGMENTATION OF AN INVARIANT INTEGRAL

Continue with the general theory. $t \to \phi_t$ is a flow on a manifold X, $t \to A_t$ is its (Eulerian) infinitesimal generator.

Let ω be an n-th degree differential form such that

$$A_t(\omega) = 0 \tag{5.1}$$

for all t ,

i.e., ω is an <u>invariant integer</u> for the flow.

Let T denote the interval of real numbers, parameterized by t. Set

$$\Omega = \omega + (-1)^r (A_t \lrcorner \omega) \wedge dt \tag{5.2}$$

$$A = A_t + \frac{\partial}{\partial t} \tag{5.3}$$

Ω is an r-form on $X \times T$. It is called the <u>augmented form</u> of ω with respect to the flow.

Similarly, A is a vector field on $X \times T$. It is called the <u>augmented vector field</u> of the flow.

Now, let us compute

$$A \lrcorner \Omega = A_t \lrcorner \omega + (-1)^r \frac{\partial}{\partial t} \lrcorner ((A_t \lrcorner \omega) \wedge dt)$$

$$= A_t \lrcorner \omega + (-1)^r (-1)^{r-1} (A_t \lrcorner \omega)$$

$$= A_t \lrcorner \omega - A_t \lrcorner \omega$$

Hence, we have

$$A \lrcorner \Omega = 0 \tag{5.4}$$

Also,

$$d\Omega \ = \ d\omega \ + \ (-1)^r \ d(A_t \lrcorner \ \omega) \quad dt$$

But,

$$d(A_t \lrcorner \ \omega) \ = \ A_t(\omega) \ - \ A_t \lrcorner \ d\omega$$

$$= \ -A_t \lrcorner \ d\omega \quad , \quad \text{using (5.1)} \quad .$$

Hence,

$$d\Omega \ = \ d\omega \ + \ (-1)^{r-1}(A_t \lrcorner \ d\omega) \wedge dt \tag{5.5}$$

This means that

> $d\Omega$ is the augmented form of the
> invariant integral $d\omega$

Hence, repeating the calculation that led to (5.4),
we have

$$A \lrcorner \ d\Omega \ = \ 0 \tag{5.6}$$

(5.4) and (5.6) now assert that:

> A is a Cauchy characteristic vector
> field of the Grassman ideal of forms
> generated by Ω and $d\Omega$.

Remark. See GPS and Cartan's "Invariant Integraux" for the definition and properties of Cauchy characteristic vector fields. In general, if A is a Cauchy characteristic vector field of the differential Grassman ideal generated by a form Ω, Cartan says that Ω is an invariant form of A. This terminology is to be avoided because of the possible confusion with the term "invariant integral".

Here is an interesting geometric consequence of the Cauchy characteristic condition. Let

$$(y^1,\ldots,y^n)$$

be a basis for the functions on $X \times T$ which are conserved under the flow generated by A. (Classically, they are called integrals of motion.) Then, Ω can be written as a differential form involving the functions

$$y^1,\ldots,y^n \quad .$$

Let us now turn to study the example that led Cartan to this general definition, the example of Hamiltonian systems

6.   THE AUGMENTED FORM ASSOCIATED WITH HAMILTONIAN SYSTEMS: THE ENERGY-MOMENTUM FORM

Let (q^i), $1 \leq i,j \leq n$, be coordinates of a configuration space of a mechanical system. Let

$$(p_i)$$

be the corresponding dual <u>momentum coordinates</u>.

Let

$$H(q,p,t)$$

be a <u>Hamiltonian function</u>. Consider Hamilton's equations:

$$\frac{dq^i}{dt} = \frac{\partial H}{\partial p_i}\,(q(t),\ p(t),\ t)$$

$$\text{(6.1)}$$

$$\frac{dp_i}{dt} = -\frac{\partial H}{\partial q^i}\,(q(t),\ p(t),\ t)$$

Let $t \to A_t$ be the infinitesimal generator of the flow associated with the differential equation (6.1). Here is the formula for it in these coordinates:

$$A_t = \frac{\partial H}{\partial p_i}\frac{\partial}{\partial q^i} - \frac{\partial H}{\partial q^i}\frac{\partial}{\partial p_i} \qquad \text{(6.2)}$$

Let

$$\omega = dp_i \wedge dq^i \qquad\qquad\qquad \text{(6.3)}$$

Then, using (6.2), (6.3),

$$A_t \lrcorner\ \omega = A_t(p_i)dq^i - dp_i A_t(q^i)$$

$$= \frac{\partial H}{\partial q^i} dq^i + \frac{\partial H}{\partial p_i} dp_i$$

$$= dH_t \quad , \tag{6.4}$$

where, for <u>fixed</u> t, H_t is the function

$$(p,q) \rightarrow H(p,q,t)$$

on R^{2n} .

Hence,

$$d(A_t \lrcorner \omega) = 0 \quad .$$

Also,

$$d\omega = 0 \quad .$$

Hence,

$$A_t(\omega) = A_t \lrcorner d\omega + d(A_t \lrcorner \omega)$$

$$= 0 \quad .$$

We see that:

> ω is an invariant integral for the flow generated by $t \rightarrow A_t$.

Remark. This is where the theory of "invariant integrals"
began, in the work of Poincaré. See his treatise "New
Methods in Celestial Mechanics", particularly Volume II.
Since the flow leaves invariant ω, it also leaves invariant

$$\omega^n \equiv \underbrace{\omega \wedge \ldots \wedge \omega}_{n \text{ copies}}$$

But,

$$\omega^n = dp_1 \wedge \ldots \wedge dp_n \wedge dq^1 \wedge \ldots \wedge dq^n \quad ,$$

which is the volume element form on R^{2n}. The invariance
of the form under the Hamiltonian flow is called Liouville's
Theorem in mechanics. It plays a key role in Statistical
Mechanics, e.g., in the work of Gibbs.

Now, let us compute the augmented form

$$\Omega = \omega - (A_t \lrcorner \omega) \wedge dt \tag{6.5}$$

$$= \text{, using (6.4)},$$

$$\omega - dH_t \wedge dt$$

$$= d(p_i dq^i - H_t dt) \tag{6.6}$$

Set:

$$\theta = p_i dq^i - H_t dt \quad . \tag{6.7}$$

θ is called the energy-momentum form. (If the Hamiltonian arises from a Lagrangian, θ is also called the Cartan form associated with the Lagrangian.) We have:

$$d\theta = \Omega \ . \tag{6.8}$$

Putting all this together, we have:

Theorem 6.1. Let H be a Hamiltonian function, and let θ be the energy-momentum form, defined by formula (6.8). Then, the solutions of the Hamiltonian equations (6.5) considered as curves in

(configuration × momentum × time)space

are characteristic curves of the exterior derivative of dθ.

This result is the main link between the traditional treatment of classical mechanics and the calculus of variations and Cartan's. Cartan emphasizes (in"Lécons sur les invariants intégraux") that assigning the two-form Ω to the Hamiltonian system brings out the true "geometrical" nature of classical mechanics.

7.   THE CONTINUITY EQUATION OF HYDRODYNAMICS AND TIME-
      DEPENDENT INVARIANT INTEGRALS

Let X be a manifold, T a time interval of real numbers. Let

$$t \rightarrow \phi_t$$

be a one-parameter family of diffeomorphisms of X. Let
$t \rightarrow \omega_t$ be a one-parameter family of volume element differential forms on X. Let

$$t \rightarrow A_t \quad \varepsilon \quad V(X)$$

be the Eulerian infinitesimal generator of the flow $t \rightarrow \phi_t$.

If O is an open subset of X, think of the real
number

$$m_t(O) \quad = \quad \int_O \omega_t$$

as the "mass" of the fluid in region O at time t.

Think of $t \rightarrow \phi_t$ as the invariant of a "fluid". If a
particle is at x at t = 0, it is at

$$\phi_t(x)$$

at time t. Let us express "conservation of mass" of the
fluid in the flow:

$$\int_O \omega_0 \quad = \quad \int_{\phi_t(O)} \omega_t \qquad (7.1)$$

(7.1) says that the mass of the fluid in O at time t = 0
is equal to the mass of fluid in the transformed region,
$\phi_t(O)$, at time t.

We can now express the mapping

$$\phi_t: 0 \to \phi_t(0)$$

as a "change of variable" in the integral (7.1):

$$\int_{\phi_t(0)} \omega_t = \int_0 \phi_t^*(\omega_t) \tag{7.2}$$

Equating (7.2) and (7.1), we have:

$$\int_0 \phi_t^*(\omega_t) = \int_0 \omega_0 \tag{7.3}$$

for <u>all</u> regions 0 in X. This condition forces the differential form equation:

$$\phi_t^*(\omega_t) = \omega_0 \quad .$$

In particular,

$$\frac{\partial}{\partial t} (\phi_t^*(\omega_t)) = 0 \quad . \tag{7.4}$$

Use the formula (2.8) defining A_t:

$$0 = \phi_t^*(A_t(\omega_t)) + \phi_t^* \left(\frac{\partial \omega_t}{\partial t} \right) \quad ,$$

or:

$$\frac{\partial \omega_t}{\partial t} + A_t(\omega_t) = 0 \qquad (7.5)$$

This is the classical <u>equation of continuity</u>, expressed in coordinate-free form.

<u>Remark</u>. We can, of course, consider Equation (7.5) as a time dependent form $t \to \omega_t$ of arbitrary degree. Such an object, satisfying (7.5), might be called a <u>time dependent invariant integral</u>.

For example, suppose that:

$$X = R^3 \quad ,$$

with (x^i), $1 \le i,j \le 3$ denoting the usual Euclidean variables Set:

$$A_t = v_t^i(x) \frac{\partial}{\partial x^i} \qquad (7.6)$$

The (v_t^i) are the <u>Eulerian velocity coordinates</u>, i.e., the velocity of the "particle" which is, at time t, at the point x.

Suppose that:

$$\omega_t = \phi_t dx^1 \wedge dx^2 \wedge dx^3 \qquad (7.7)$$

ϕ_t is called the <u>density function</u>.

<u>Exercise</u>. Show that Equation (7.5) takes the following form:

$$\frac{\partial}{\partial t} \phi_t + \frac{\partial}{\partial x^i} (\phi_t v^i) = 0 \tag{7.8}$$

Equation (7.8) is classically called the <u>continuity equation</u>
<u>of fluid flow</u>. It is one of the basic equations. (See
Volume V of IM.)

   We can define the augmented form for an arbitrary time-
dependent invariant integral:

$$\Omega = \omega_t - dt \wedge (A_t \lrcorner \omega_t) \tag{7.9}$$

Again, set:

$$A = A_t + \frac{\partial}{\partial t} \tag{7.10}$$

A and Ω are then geometric objects defined on

$$X \times T .$$

By the way we have constructed them, we have:

$$A \lrcorner \Omega = 0 . \tag{7.11}$$

Also,

$$d\Omega = d\omega_t + dt \wedge \frac{\partial \omega_t}{\partial t} + dt \wedge (d(A_t \lrcorner \omega_t))$$

$$= d\omega_t + dt \wedge \frac{\partial \omega_t}{\partial t} + dt \wedge (A_t(\omega_t) - A_t \lrcorner d\omega_t)$$

$$= \quad , \text{ using } (7.5),$$

$$d\omega_t - dt \wedge (A_t \lrcorner \; d\omega_t) \qquad\qquad (7.12)$$

Remark. $d\omega_t$ denotes the exterior derivative of the form, with t held as a fixed parameter.

Let us specialize these formulas to the case of interest in hydrodynamics, with ω_t given by (7.7), A_t by (7.7). Then,

$$\Omega = \rho_t dx^1 \wedge dx^2 \wedge dx^3 - dt \wedge \rho_t (v_t^1 dx^2 \wedge dx^3 - v_t^2 dx^1 \wedge dx^3$$

$$+ v_t^3 dx^1 \wedge dx^2)$$

$$(7.13)$$

This form is called the energy-momentum form of the fluid flow.

Exercise. Show that the equation of continuity, (7.8), is equivalent to the condition

$$d\Omega = 0 \qquad\qquad (7.14)$$

8.   CARTAN'S EQUATIONS OF CONTINUUM MECHANICS MOTION

In his magnificent paper "Sur les variétés à connexion affine et la theorie de la relativité genéralisée", Cartan described an amazingly simple way to write down the equations of fluid flow motion. This work seems to be completely unknown today.

Let X continue to denote R^3, with Cartesian coordinates (x^i), $1 \leq i,j \leq 3$. Let T denote a time interval, with parameter denoted by t.

Let $t \to \phi_t$ be a flow on X, which represents the actual fluid flow. Let

$$A_t = v_t^i(x) \frac{\partial}{\partial x^i} \tag{8.1}$$

be the Eulerian infinitesimal generator: physically, it is the _velocity field_. Let $\rho_t(x)$ denote the density function.

Using this data, construct the following differential forms:

$$\omega_t = \rho_t \, dx \tag{8.2}$$

$$\omega_t^i = v^i \rho_t \, dx \quad , \tag{8.3}$$

with

$$dx = dx^1 \wedge dx^2 \wedge dx^3 \quad . \tag{8.4}$$

As we have seen in Section 7, ω_t defines the _mass_ of the fluid flow, i.e., for any region O in X,

$$\int_0 \omega_t$$

is the total mass of the fluid which is in O at time t.

Similarly,

$$\left(\int_0 \omega_t^i \right) \frac{\partial}{\partial x^i} \tag{8.5}$$

is the <u>momentum vector</u> of the fluid which is in the region O at time t.

In order to express Newton's dynamical equations (i.e., rate of change of momentum = force) in an elegant geometric way, Cartan next constructs the augmented forms:

$$\Omega = \omega_t - dt \wedge (A_t \lrcorner \omega_t) \tag{9.6}$$

$$\Omega^i = \omega_t^i - dt \wedge (A_t \lrcorner \omega^i) \tag{8.7}$$

Next, he postulates three differential forms of the following form:

$$\Pi^i = -p^{ij} \left(\frac{\partial}{\partial x^j} \lrcorner dx \right) \wedge dt \tag{8.8}$$

They are called the <u>pressure forms</u>. The components p^{ij} depend, respectively, on i and j.

Finally, there are forms

$$\Sigma^i = F^i dt \wedge dx \quad , \tag{8.9}$$

which define the <u>exterior forces</u>.

The equations of motion of the fluid are now:

$$d\Omega = 0 \tag{8.10}$$

$$d(\Omega^i + \Pi^i) = \Sigma^i \tag{8.11}$$

<u>Exercise</u>. Show that (8.10) and (8.11) are equivalent to the following equations:

$$\frac{\partial \rho_t}{\partial t} + \frac{\partial}{\partial x^i} (\rho_t v^i) = 0 \tag{8.12}$$

$$\rho_t \frac{\partial v^i}{\partial t} + v^j \frac{\partial v^i}{\partial x^j} + \frac{\partial}{\partial x^j} (p^{ij}) = F^i \tag{8.13}$$

In these equations, the v^i, ρ, p^{ij} are the unknowns There are thus

$$3 + 1 + 16 = \underline{10}$$

unknowns. There are four equations. To make a determined set, one must add "constitutive relations" which relate the pressure tensors p^{ij} to the kinematic data. For example, in the case of a simple <u>fluid</u>,

$$p^{ij} = p\delta^{ij} ,$$

i.e., the "pressure tensor" is characterized by a scalar
quantity. A typical constitutive relation in this case involves
a functional relation between the density ρ and the pressure p.

9. RELATIVISTIC CONTINUUM MECHANICS EQUATIONS AND FURTHER GENERALIZATIONS

An enormous number of papers have been written in the
last twenty years on the general form of continuum mechanics
equations. This has involved two distinct groups of people--
physicists and astrophysicists interested in relativistic
continuum mechanics, and applied mathematicians interested
in classical continuum mechanics. Cartan's method gives an
amazingly simple approach to both areas!

Let Y be a four-dimensional manifold. (A special case
will be the situation $Y = X \times T = R^3 \times R \equiv R^4$ considered in
previous sections.)

Choose indices as follows, and the summation convention
on these induces:

$$0 \leq \mu,\nu \leq 3$$

Let

$$(y^\mu)$$

be a coordinate system for X'. Set

$$dy \;=\; dy^0 \wedge dy^1 \wedge dy^2 \wedge dy^3 \qquad\qquad (9.1)$$

Consider a vector field

$$A \;=\; A^\mu \frac{\partial}{\partial y^\mu}$$

on   Y. Set:

$$\Omega^\mu \;=\; \rho A^\mu (A \;\lrcorner\; \cdot dy)$$

We can also postulate forms

$$\Pi^\mu$$

and four-forms Σ^μ. The equations of motion are then:

$$d(\Omega^\mu + \Pi^\mu) \;=\; \Sigma^\mu \quad. \qquad\qquad (9.1)$$

We can now write this in a completely coordinate free
way. Set:

$$\underline{\Omega} \;=\; \frac{\partial}{\partial x^\mu} \otimes \Omega^\mu$$

$$(9.2)$$

$$\equiv\; \rho A \otimes (A \;\lrcorner\; dy)$$

$\underline{\Omega}$ is a <u>differential form on</u>   Y   <u>with values in the tangent</u>
<u>bundle</u>. Similarly, define $\underline{\Pi}$ and $\underline{\Sigma}$ as tangent-vector valu

differential forms on Y, and write (9.1) as:

$$d(\underline{\Omega} + \underline{\Pi}) = \underline{\Sigma} \tag{9.3}$$

To define the "d" operation needed for (9.3), we must
postulate an affine connection on Y, i.e., a covariant
derivative operation

$$(A,B) \rightarrow \nabla_A B$$

for vector fields on Y. "d" is then defined by the usual
formula for exterior derivatives. For example, if $\underline{\theta}$ is a
tangent-vector valued one-form,

$$d\underline{\theta}(A,B) = \nabla_A(\underline{\theta}(B)) - \nabla_B(\underline{\theta}(A)) - \underline{\theta}([A,B]) \tag{9.4}$$

for $A, B \in V(Y)$

Of course, further differential geometric structure is
implicitly introduced into Y by the constitutive relations
which one must add to (9.3) in order to define a deterministic
system.

In summary, the main qualitative point one sees from a
proper understanding of Cartan's brief work on continuum mechanics is that the basic objects should be defined in terms of
tangent vector-valued differential forms. An intensive study
of the whole spectrum of continuum mechanics from this point
of view is obviously called for.

Chapter 13

OPTIMIZATION AND PROGRAMMING OF
FUNCTIONS ON MANIFOLDS

1. INTRODUCTION

In this chapter, I have several goals in mind. The
most obvious one is to develop the classical material concerning the applications of differential calculus to the
problem of optimizing a real-valued function on Euclidean
space or a manifold, with and without constraints. However,
this material is traditionally in close relation to mathematical economics. See "Foundations of Economic Analysis",
by P. A. Samuelson, and the book by Intriligator [1].

As Samuelson himself remarks, in the book quoted above
and in several papers in Volume I of his Collected Works,
there are strong analogies between classical mathematical
economics and equilibrium thermodynamics, particularly as
developed by J. W. Gibbs. Understanding these analogies
will, I believe lead to development of a general "systems
theory," which will reach into physics and economics. In
addition to the classical "equilibrium" theory, there are
also interesting economics applications of optimal control
theory on the horizon.

2. EXTREME POINTS OF FUNCTIONS AND CALCULUS

In this section, I shall review the traditional way that differential calculus is used to study maxima and minima of real-valued functions defined on Euclidean spaces.

Choose indices varying over the following range, with the summation convention:

$$1 \leq i, j \leq n.$$

$x = (x_i)$, $y = (y_i)$, denote points of R^n. Let

$$f: R^n \to R$$

be a real valued function assumed, for simplicity, to be C^∞ and everywhere defined.

Given a point $y \in R^n$, f can be expanded in a <u>Taylor series</u> about y:

$$f(x) \sim f(y) + a_i(x_i - y_i)$$
$$+ a_{ij}(x_i - y_i)(x_j - y_j) \qquad (2.1)$$
$$+ \ldots$$

(The coefficients a_i, a_{ij},... depend on y. Of course they are just the partial derivatives of f at y). In calculus, one learns how to write this as an <u>equality</u>, with a remainder after n terms. We shall not need to know the explicit formulas for this remainder, only that it <u>exists</u>.

For example, here is Taylor's Theorem, with remainder after
quadratic terms. (The higher order remainder form is not
very useful, due to algebraic difficulties with polynomials
of degree higher than two.)

<u>Theorem 2.1</u>. There are C^{∞} functions $a_i(y)$, $a_{ij}(y)$, $a_{ijk}(x,y)$
such that:

$$f(x) = f(y) + a_i(y)(x_i - y_i)$$
$$+ a_{ij}(y)(x_i - y_i)(x_j - y_j) \qquad (2.2)$$
$$+ a_{ijk}(x,y)(x_i - y_i)(x_j - y_2)(x_k - y_k).$$

<u>Exercise</u>. Show that the functions $y \to (a_i(y), a_{ij}(y))$ which
appear in (2.2) are uniquely determined by f, and are given
by the classical formulas:

$$a_i = \frac{\partial f}{\partial x_i}$$
$$a_{ij} = \frac{1}{2} \frac{\partial^2 f}{\partial x_i \partial x_j}. \qquad (2.3)$$

With the aid of (2.2), let us study the behavior of f
in a neighborhood of a given point $y \in R^n$.

<u>Definition</u>. f has an <u>extreme point</u> at y if:

$$a_i(y) = 0. \qquad (2.4)$$

y is a <u>relative maximum</u> (<u>minimum</u>) of f if:

$$f(x) - f(y) \leq 0 \ (\geq 0) \qquad\qquad (2.5)$$

for all x sufficiently close to y.

Here is the usual picture of the graph of f (in case n = 1) in the case of a relative maximum and minimum:

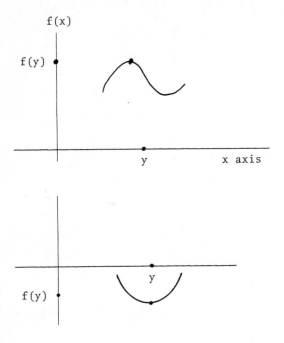

Theorem 2.2. If y is a relative maximum or minimum of f, then y is an extreme point of f.

Proof. Assume (2.3), i.e. the coefficients a_i in the Taylor expansion of f are given by the first partial derivatives of f. We must show that (2.5) implies

$$a_i(y) = 0.$$

For example, consider a_1. Here is its definition in terms of limits:

$$a_1(y) = \lim_{\varepsilon \to 0} \frac{f(y + (\varepsilon, 0)) - f(y)}{\varepsilon}. \tag{2.6}$$

Suppose, for example, that y is relative minimum. Then, the right hand side of (2.6) is ≥ 0, for ε sufficiently small, whether ε is positive or negative. Hence,

$$a_1(y) \geq 0. \tag{2.7}$$

Change ε to $-\varepsilon$ in the right hand side of (2.7)

$$a_1(y) = \lim_{\varepsilon \to 0} - \left(\frac{f(y + (-\varepsilon, 0)) - f(y)}{\varepsilon} \right). \tag{2.8}$$

The term in parentheses on the right hand side of (2.8) is again ≥ 0, since y is a relative minimum. Hence,

$$a_1(y) \leq 0 \tag{2.9}$$

(2.6) and (2.8) together imply:

$$a_1(y) = 0.$$

The argument can be repeated for all i, to prove (2.4).

Remark: Of course, a point y can be an extreme point, but not a maxima or minima. For example,

$$n = 2, \ y = 0,$$
$$f(x) = x_1^{\ 2} - x_2^{\ 2}.$$

Such an extreme point is often called a <u>saddle point</u>. Another typical situation is an <u>inflection point</u>, e.g.

$$n = 1, \ y = 0,$$
$$f(x) = x^3$$

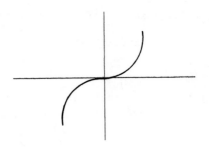

Here are the standard necessary conditions on the second derivatives that a point y be a relative minimum or maximum.

<u>Theorem 2.3</u>. Suppose y is a relative minimum of f. Then, the quadratic form on R^n whose matrix is $(a_{ij}(y))$ is <u>positive semi-definite</u>, i.e.

$$a_{ij}(y)x_i x_j \geq 0 \qquad\qquad (2.10)$$
$$\text{for all } x \in R^n.$$

Exercise. Prove Theorem (2.3).

Of course, if y is a relative maximum of f, it is a relative minimum of -f, hence a necessary condition that y be a relative maximum is that the quadratic form be negative semi-definite.

Here is the classical sufficient condition.

Theorem 2.4. Suppose that y is an extreme point of f, and that is quadratic form $(a_{ij}(y))$ is positive definite, i.e.

$$a_{ij}(y)x_i x_j > 0 \qquad (2.11)$$

$$\text{for all } x = (x_i) \in R^n - (0).$$

Then,

$$f(y) < f(x) \qquad (2.12)$$

for all x sufficiently close, but unequal, to x. In particular, y is a relative minimum of f.

Remark: If y is an extreme point, and if (2.11) is satisfied, then y is called a strong relative minimum.

Let us now formulate an analogous concept for extreme points.

Definition. An extreme point y of f is said to be a nondegenerate extreme point of f if the determinant of the matrix

$$\frac{1}{2} \, (a_{ij}(y)) \tag{2.13}$$

is different from zero.

Remarks: The matrix (2.13) is called the <u>Hessian of f at y</u>.
It is, of course, given by the second partial derivatives
of f:

$$\left(\frac{\partial^2 f}{\partial x_i \partial x_j} \, (y) \right)$$

Here is a diagram which illustrates the logical interconnection between these notions.

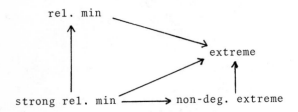

(non-deg. extreme) $\cup$ (rel. min) $\longrightarrow$ strong rel. min.

Finally, here is a result which shows more precisely
why the quadratic forms are the key examples for the study
of non-degenerate extreme points.

Theorem 2.5.   (Marston Morse [1]).   Suppose that y is a

non-degenerate extreme point of f. Then, there is a new
coordinate system of R, valid in some neighborhood of y,
such that $x \to (f(x) - f(y))$ is a quadratic form in these
new variables.

Exercise. Prove Theorem 2.5.

Exercise. If y is a non-degenerate extreme point of f, show
that there is a neighborhood of y in R^3 which contains no
other extreme points, i.e. a non-degenerate extreme point
of a C^∞ (or even a C^2) function is an isolated extreme point.

    Now we formulate these ideas for functions on manifolds,
in a coordinate-free way. This formulation is essentially
trivial, but will be useful later on. The main point is
that we want to have an "intrinsic" definition of the Hessian
form associated to a function at an extreme point.

3. EXTREME OF FUNCTIONS ON MANIFOLDS. MORSE THEORY

    Let M be a manifold, and let

$$f: M \to R$$

be a real-valued, C^∞ function on M. In order to formulate
the notion of "extreme point," we recall some basic differential-

geometric notions.

Let $T(M)$ denote the tangent vector bundle to M. For
$p \in M$, the fiber of this bundle is the tangent vector space

$$M_p$$

to M at p. Recall that each $v \in M_p$ is an R-linear map

$$F(M) \to R$$

such that:

$$v(f_1 f_2) = v(f_1) f_2(p) + v(f_2) f_1(p)$$

$$\text{for } f_1, f_2 \in F(M).$$

(3.1)

Each vector field $X \in V(M)$ (defined as a derivation of the
algebra $F(M)$) determines a cross-section $p \to X(p)$ of $T(M)$:

$$X(p)(f) = X(f)(p).$$

In this way,

$$V(M) = \Gamma(T(M)),$$

i.e. the space of vector fields is identified with the space
of cross-sections of the vector bundle. Recall also that,
for

$$M = R^n,$$

each vector field X is a first order, linear differential
operator:

$$X = A_1 \frac{\partial}{\partial x_1} + \ldots + A_n \frac{\partial}{\partial x_n}.$$

The exterior derivative df is then a linear map

$$df: T(M) \to R.$$

Explicitly,

$$df(v) = v(f)$$

for $v \in T(M)$.

For $p \in M$, $M_p^{\ d}$ denotes the dual space to M_p. The elements of M_p are called <u>1-covectors of M at p</u>. As p varies, $M_p^{\ d}$ is the fiber of a vector bundle $T^d(M)$, called the <u>cotangent bundle to M</u>. df then defines a element of $\Gamma(T^d(M))$, the space of its cross-sections. For $p \in M$, df(p) denotes the <u>value</u> of the cross-section at p, which is of course an element of $M_p^{\ d}$.

Using these notions, we can define the notion of "extreme point" for a function on M, in such a way that it reduces to the classical notion (described in Section 2) when $M = R^n$.

<u>Definition</u>. A point $p \in M$ is an <u>extreme point</u> of f if:

$$df(p) = 0. \qquad\qquad (3.2)$$

The notion of "relative minimum and maximum" defined for functions on R^n in Section 2 of course carries over to

manifolds (and even general topological spaces). For
example:

Definition. A point p ∈ M is a relative minimum for f if
there is an open subset U,

    p ∈ U ⊂ M,

such that:

    f(q) ≥ f(p)   (3.3)

    for all q ∈ U.

If p is a relative minimum for -f, then p is a relative
maximum for f.

Exercise. If p is a relative minimum or maximum for f, show
that:

    df(p) = 0,

i.e. p is an extreme point for f.

In Section 2, we have seen that the type of extremal,
i.e. minima, maxima, saddle or inflection point, of a
function on R^n is related to the matrix of second partial
derivatives. We shall now formulate this in a coordinate-
free way.

Theorem 3.1. Let p be an extreme point of the function f: $M \to R$, i.e. a point $p \in M$ such that

$$df(p) = 0. \tag{3.4}$$

Then, there is a symmetric, bilinear form

$$h_f: M_p \times M_p \to R,$$

called the Hessian of f at p, such that:

$$h_f(X(p), Y(p)) = X(Y(f))(p) \tag{3.5}$$

for X, Y $\in$ V(M).

Proof. Consider the mapping

$$(X, Y) \to X(Y(f)(p) \equiv \alpha(X, Y). \tag{3.6}$$

of $V(M) \times V(M) \to R$. If $X(p) = 0$, then $\alpha(X, Y) = 0$. We shall prove that:

$$\text{If} \quad Y(p) = 0, \text{ then } \alpha(X, Y) = 0. \tag{3.7}$$

To prove (3.7), suppose Y is of the special form:

$$Y = gZ, \tag{3.8}$$

with $g \in F(M)$, $Z \in V(M)$, such that:

$$g(p) = 0. \tag{3.9}$$

Then,

$$\alpha(X, Y)(p) = X(gZ)(f)(p)$$
$$= X(g)Z(f)(p) + g(p)XZ(f)(p),$$

which is zero because of (3.4) and (3.9). But, an arbitrary
vector field Y which vanishes at p is the sum of vector
fields of the form (3.8)-(3.9), which proves (3.7).

Now, M_p is the quotient of the vector space V(M) by the
linear subspace of those which vanish at p. As we have just
seen, α then passes to the quotient to define a bilinear
form

$$M_p \times M_p \to R.$$

Let us <u>define</u> h_f as this form α, passed to the quotient.
Following the logic, (3.5) is automatically satisfied. The
symmetry of h_f follows from the symmetry of α:

$$\begin{aligned}
\alpha(X, Y) &= \alpha(Y, X) \\
&= (X, Y(f)) - Y(X(f))(p) \\
&= [X, Y](f)(p) \\
&= 0, \text{ since } [X, y] \in V(M), \text{ and} \\
&\quad (3.4) \text{ is satisfied.}
\end{aligned}$$

Remark. Local Coordinates

We can now see the connection between the <u>Hessian matrix</u>
of a function on R^n, defined in Section 2, and the Hessian
<u>form</u> h_f. Let (x_i), $1 \le i, j \le n$, be a local coordinate
system for M. Suppose that:

$$X = A_i \frac{\partial}{\partial x_i}$$

$$Y = B_i \frac{\partial}{\partial x_i},$$

with A_i, $B_i \in F(M)$.

Exercise. Show that

$$h_f(X(p), Y(p)) = \frac{\partial^2 f}{\partial x_i \partial x_j}(p) A_i(p) B_j(p). \qquad (3.10)$$

Exercise. If p is a relative minimum for p, show that:

$$h_f(v, v) \geq 0 \qquad (3.11)$$

for all $v \in M_p$

i.e. h_f is a non-negative form.

Using the algebraic property of the Hessian form, we can now formulate some basic ideas. They arise from Morse Theory (see Morse [1], Milnor [1]), which is usually considered to be a part of Differential Topology, but we shall see that they are useful in applied optimization problems also.

Definition. p is a strong relative minimum of f if

$$h_f(v, v) > 0 \qquad (3.12)$$

for $v \in M_p - (0)$,

i.e. if the Hessian form h_f if positive definite.

The algebraic properties of h_f are also important for the study of the extremal points of f which are not maxima or minima, e.g. which are of "saddle point" type. Here are some relevant concepts.

Definition. Let p be an extreme point of f, and let h_f be its Hessian form. Set:

$$N_f = \{v \in M_p : h_f(v, M_p) = 0\}. \qquad (3.13)$$

The vectors on N_f are called the null-vectors of f. The dimension of N_f is called the nullity of the extreme point. p is called a non-degenerate extreme point if its nullity is zero, i.e. if h_f is a non-degenerate form.

We can now assign an integer to each non-degenerate extreme point.

Definition. Let p be a non-degenerate extreme point of f. The Morse index of f is the integer n satisfying the following conditions:

> There is a linear subspace $\gamma \subset M_p$
> of dimension n such that h_f re-   (3.14)
> stricted to γ is negative definite.

> n is the largest integer satisfying
> (3.14), i.e. if γ' is a linear sub-
> space of M_p of dimension n+1, then   (3.15)
> h_f restricted to γ' is not negative.

Remark: Using facts about the algebra of real quadratic forms (which are easy to prove) this awkward definition may be put in a more elegant form. n is the _Morse index_ if there are linear subspaces γ, $\gamma' \subset M_p$ such that:

$$M_p = \gamma \oplus \gamma'$$

$$h_f(\gamma, \gamma') = 0$$

$$h_f(v, v) < 0 \text{ for } v \in \gamma - (0)$$

$$h_f(v, v) > 0 \text{ for } v \in \gamma' - (0)$$

$$\dim \gamma = n.$$

Example. $M = R^m$, $p = (0)$.

$$f(x) = \frac{1}{2}(-x_1{}^2 - \ldots - x_n{}^2 + x_{n+1}{}^2 + \ldots + x_m{}^2). \quad (3.15)$$

Then, the point 0 is a critical point of f. The Hessian is readily computed:

$$\frac{\partial^2 f}{\partial x_i \partial x_j} = \begin{cases} - \delta_{ij} \text{ for } 1 \le i, j \le n \\ 0 \text{ for } 1 \le i \le n, n + 1 \le j \le m \\ - \delta_{ij} \text{ for} \end{cases}$$

We see from this formula that $M_p \equiv R_0{}^m$, which is in turn isomorphic to R^m itself, is the direct sum of the subspace spanned by $\frac{\partial}{\partial x_1}, \ldots, \frac{\partial}{\partial x_n}$, on which h_f is negative definite, and the subspace spanned by $\frac{\partial}{\partial x_{n+1}}, \ldots, \frac{\partial}{\partial x_m}$, on which h_f is positive definite. We see then that the Morse index of this function is n.

If $m = 2$, $n = 1$, the graph of f represents a <u>saddle point</u> (or <u>col</u>), e.g. between two mountains. The fact that the Hessian is in certain directions negative, in certain positive, means that f decreases in certain directions leading away from the point, and increases in other directions. This corresponds, of course, with what mountain climbers or map-makers mean.

Return to a general manifold M. Here is a result which explains more precisely the geometric meaning of the way the Hessian h_f and the Morse index measure the "saddle point" property.

<u>Theorem 3.2</u>. Let $f: M \to R$ be a function with an extreme point at p. Let $t \to \sigma(t)$, $-1 < t < 1$, be a curve in M such that:

$$\sigma(0) = p$$
$$\sigma'(0) = v \in M_p,$$

i.e. v is the tangent vector to M at p. Then,

$$\frac{d}{dt} f(\sigma(t))/_{t=0} = 0 \qquad\qquad (3.17)$$

$$\frac{d^2}{dt^2} f(\sigma(t)) = h_f(v, v).$$ (3.18)

In particular, if $h_f(v, v) > 0$, then $t \to f(\sigma(t))$ has a strong relative minimum at $t = 0$, while if $h_f(v, v) < 0$, $t \to f(t)$ has a strong relative maximum. In particular, we see that the Morse index gives the maximal number of independent directions going out from p in which the function increases.

Remark: Morse Theory (see Morse [1] and Milnor [1]) deals with the interrelation between the number of extreme points of given index and the topological properties of a manifold M. Here are the Morse inequalities (formula (3.19), which is the most famous result:

> Let $f: M \to R$ be a function on a manifold M which has only non-degenerate extreme points. For each integer n, let m(n) be the number of critical points of f which have n as Morse index. Let b(n) be the n-th Betti number of M, i.e. the dimension of the n-th homology group of M (with any field as coefficients). Then,
>
> $m(n) \geq b(n).$ (3.19)

Thus, the Morse inequalities assert the <u>existence</u> of
extreme points of given index n, if b(n) is positive.
Intuitively, the topological "richness" of M (indicated by
the presence of homology groups, i.e., "holes" in the space)
<u>requires</u> f to have certain saddle point type of extreme
points. This type of result obviously might be of enormous
significance in applications, e.g., to economics, but this
mainly lies in the future. (See Dieᴿcker [1] for an interesting application of differential-topological ideas to economics.
This might be the "camel's nose in the tent", as far as the
application of topology to economics goes!)

The simplest interesting example is the two-dimensional
manifold called the <u>torus</u>. Laymen know this as the <u>doughnut</u>

Its Betti-numbers are:

$$b(0) \;=\; 1 \;=\; b(1) \;=\; 1 \;=\; b(2) \quad .$$

In particular, any real valued function on M (with only
non-degenerate extreme points) <u>must</u> have a saddle-point extrema

The inequalities (3.19) were proved by Morse in the
1930's. Research by topologists since 1950, e.g., by Bott,
Milnor, Smale and many others, has been much concerned with
this topic. Techniques are developed for relating "finer"

topological invariants of M than the Betti numbers, for seeing when the inequalities (3.19) become <u>equalities</u>, and, to a certain extent, generalizing to the case where f has degenerate extrema. Unfortunately, this work has not been in close touch with applications (which are the natural source of functions with complicated and interesting extremal properties) and has become somewhat sterile and precious. In parallel, and curiously with no contact with the "pure" work, there has been a vast development of what is called "programming" in the applied literature, which is basically the same mathematical topic. Presumably, future historians of science will be puzzled by this lack of contact! Establishing such contact is what I regard as "Interdisciplinary Mathematics".

4.   PERTURBATION OF NON-DEGENERATE EXTREMALS

As Samuelson points out [1], many problems in mathematical economics involve perturbation of extrema. Here is his way of describing the problem. Suppose that:

$$x = (x_1, \ldots, x_n) \in R^n$$

α is a real variable

$$f(x, \alpha)$$

is a function of these variables. Suppose that, for fixed
α_0, x^0, is a point of R^n which is a minimum of the function

$$f_{\alpha_0} : x \rightarrow f(x, \alpha_0).$$

What happens to this minimum as α varies? Is there a function

$$\alpha \rightarrow x(\alpha)$$

of $R \rightarrow R^n$ such that:

x(α) is the minimum of $x \rightarrow f(x, \alpha)$? For example,
this minimal $x(\alpha)$ may be "prices" or "production quantities,"
which vary in response to a change in α, e.g. α the "tax
rate" or "interest rate," etc. Thus, one is particularly
interested in

$$\frac{dx}{d\alpha},$$

the rate of change of the minimal x with respect to change
in α. (Note that economists traditionally use the term
"marginal" to denote "partial derivative" or "rate of
change.")

Often this "marginal" quantity can be calculated using
calculus. The condition that $x(\alpha)$ is extremal, for fixed
α, is that

$$\frac{\partial f}{\partial x_i} (x(\alpha), \alpha) = 0.$$

This can be differentiated with respect to :

$$\frac{\partial^2 f}{\partial x_i \partial x_j} (x(\alpha),\alpha) \frac{dx_j}{d\alpha} + \frac{\partial^2 f}{\partial \alpha \partial x_i} (x(\alpha),\alpha) = 0$$

Thus, $dx/d\alpha$ can be calculated <u>if</u> the Hessian $(\partial^2 f/\partial x_i \partial x_j)$ is a non-singular matrix. Now, one is particularly interested in the conditions that certain of the "marginal" quantities $dx_i/d\alpha$ have fixed sign, since this often has an important meaning. For example, in which direction should prices change in response to a change in the tax rate? Samuelson, in his book [1] and in many of his papers of Volume 1 of his Collected Works, provides an extensive discussion of these matters.

One of my major aims in this volume is to understand these perturbation problems from the point of view of mani- fold theory. In this I shall proceed in a relatively simple- minded way, using the techniques of manifold theory, particu- larly the Implicit Function Theorem.

Here is a basic set-up. Let M, B be manifolds, and let

$$\pi : M \to B$$

be a map. For b ε B, let

$$M(b) \equiv \pi^{-1}(b) = \{p \in M: \pi(p) = b\}$$

denote the fiber of π, over the point $b \in B$. Here is the
picture describing the fibers.

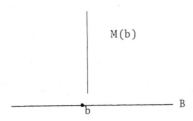

Let us <u>assume</u> that each subset

$$M(b)$$

can be made into a <u>submanifold</u> of M. Recall that this has
a precise technical meaning in manifold theory. Precisely,
it means that, for each $b \in B$, there is a manifold N_b and a
map

$$\varphi_b: N_b \rightarrow M$$

such that:

$$\varphi_b(N_b) = M(b)$$

$$(\varphi_b)_*: T(N_b) \rightarrow T(M)$$

is one-one.

It is customary, however, to relax this precision, and
regard M(b) as the "submanifold." In <u>local</u> problems, which

are our main concern, at least for the moment, one cannot
usually get into trouble with rigor by looking at things
in the relaxed way.

As part of the convention, for each point $p \in M(b)$,
we denote by

$$M(b)_p$$

the tangent space to $M(b)$ at p, and consider it as a linear
subspace of M_p. (Strictly speaking, this means identifying
the tangent space $(N_b)_{\varphi_b^{-1}(p)}$ with its image in M_p under the
one-one map

$$(\varphi_b)_* : (N_b)_{\varphi_b^{-1}(p)} \to M_p.$$

Now, let

$$f : M \to R$$

be a C^∞ function. For $b \in B$, let

$$f_b = f \text{ restricted to } M(b) \in F(M(b)).$$

Our problem: Study the extremal points of f_b, with particular
attention to how they vary with b.

It should be reasonably clear to the reader why this
is the differential-geometric version of the problem Samuelson has abstracted from economics problems. To see this
more explicitly, suppose $x = (x_1, \ldots, x_n) \in R^n$, $\alpha \in R$, with
$f(x, \alpha)$ a function of both variables. One can now set:

$$M = R^n \times R,$$

$$B = R$$

$\pi: M \to R$ the Cartesian projection,

$$\pi(x, \alpha) = x.$$

Return to the general situation. Let us suppose that the following condition is satisfied:

Maximal rank assumption

$\pi: M \to B$ is a <u>maximal rank mapping</u>, in the sense that:

$$\pi_*(M_p) = B_{\pi(p)} \tag{4.1}$$

for all $p \in M$.

With this assumption, here is the main result.

Theorem 4.1. Suppose that (4.1) is satisfied, with $f: M \to R$ a C^∞, real valued function. Set:

$$M' = \{p \in M: f_{\pi(p)} \text{ has a non-degenerate} \tag{4.2}$$
$$\text{extremal point at } p\}.$$

Then, M' is a submanifold of M such that the map

$$\pi: M' \to B$$

is a local diffeomorphism. In particular,

$$\dim M' = \dim B. \tag{4.3}$$

Here is an immediate consequence of Theorem 4.1, which is closer in spirit to the form used by the economists, e.g. by Samuelson [1] Chapter III.

Corollary to Theorem 4.1. For each point $b_0 \in B$, and each point $p_o \in \pi^{-1}(b_0) \equiv M(b_0)$ such that f_{b_0} has a non-degenerate extremal point at p_0, there is an open neighborhood U,

$$b \in U \subset B,$$

and a cross-section map

$$\varphi: U \to M,$$

(i.e. such that $\pi\varphi$ = identity) such that:

> For $b \in U$, $\varphi(b)$ is a non-degenerate
> extremal point of f_b.

Proof of Theorem 4.1. It will be shown that suitable coordinates may be chosen for M, so that the standard implicit function argument (as sketched in the beginning of this section) applies. Let

$$(b_1, \ldots, b_n) \equiv (b_a), \ 1 \le a, b \le n,$$

be a coordinate system of functions on B. This means, of course, that the 1-forms db_a are linearly independent. Set:

$$\alpha_a = \pi^*(b_a). \tag{4.4}$$

Exercise. Show that the maximal rank condition for π, i.e. condition (4.1) is equivalent to the following condition:

> The 1-forms on M,
>
> $$\pi^*(db_1), \ldots, \pi^*(db_n) \qquad (4.5)$$
>
> are linearly independent

Hint: A linear map between two vector spaces is onto if and only if its dual map is one-one.

Since (4.1) is a hypothesis for Theorem 4.1, we combine (4.4) and (4.5) to infer that the functions $\alpha_1, \ldots, \alpha_n$ on M are functionally independent. The Implicit Function Theorem implies (e.g. see DGCV, p. 30) that there is, locally, a set of functions

$$x = (x_1, \ldots, x_m) \equiv (x_i), \ 1 \le i, \ j \le m.$$

on M such that:

> (x, α) forms a local coordinate
> system for M.
> $\qquad (4.6)$

The α are "base-like," i.e. constant on the fibers of π, i.e. the submanifolds

> $M(b)$.

Hence, the functions $(x_1, \ldots, x_m)$ when restricted form a local coordinate system for these fibers.

In particular, the given function f becomes a function

$$f(x, \alpha)$$

of these variables. The variation of f with <u>respect to x</u> determines how the function varies along the fibers.

Now, let p_0 be a point of M,

$$b_0 = \pi(p_0),$$

such that f_{b_0} has <u>a non-degenerate extremal point at</u> p_0.
In terms of the statement (4.2) which defines M', this just means that:

$$p_0 \in M'.$$

The conditions for a "non-degenerate extremal point" are then:

$$\frac{\partial f}{\partial x_i}(p_0) = 0 \tag{4.7}$$

$$\det\left(\frac{\partial^2 f}{\partial x_i \partial x_j}(p_0)\right) \neq 0. \tag{4.8}$$

<u>To prove</u> Conditions (4.8) imply that the:

> The 1-forms $d\left(\frac{\partial f}{\partial x_i}\right)$ are linearly independent at p_0. (4.9)

<u>Proof</u>. Suppose that (4.9) were not satisfied, i.e. there would be a non-trivial relation of the form:

$$\lambda_i d(\frac{\partial f}{\partial x_i})(p_0) = 0,$$

with $\lambda_i \in R$. Then,

$$0 = \lambda_i (\frac{\partial^2 f}{\partial x_i \partial x_j} dx_j + \frac{\partial^2 f}{\partial x_i \partial \alpha_a} d\alpha_a) = 0,$$

which forces (since the dx_j, $d\alpha_a$ are linearly independent) the relation

$$\lambda_i \frac{\partial^2 f}{\partial x_i \partial x_j} = 0,$$

which contradicts (4.8).

Exercise. Does (4.9) conversely imply (4.8)?

To Prove Conditions (4.8) imply that:

The 1-forms $d(\frac{\partial f}{\partial x_i})$, $d\alpha_a$ (4.10)

are linearly independent at p_0.

Exercise. Prove (4.10).

Hint: The proof is similiar to the proof of (4.9).

   Proceed to the proof of the Theorem. By hypothesis,

$$\frac{\partial f}{\partial x_i} (p_0) = 0$$

If $U \subset M$ is an open neighborhood of p_0 in which the

$(\frac{\partial f}{\partial x_i}, \alpha_a)$ form a coordinate system (the Implicit Function
Theorem and (4.10) imply that such a neighborhood exists),
then

$\qquad$ M' ∩ U consists of the points p ∈ U at which

$\qquad$ $\frac{\partial f}{\partial x_i} = 0$.

This suffices to give the manifold structure to M'. Notice
that the functions $\alpha_a \equiv \pi^*(b_a)$ form a coordinate system for
this manifold structure. In particular (4.3) follows.
The fact that π restricted to M' is a local diffeomorphism
also follows.

Remark: Recall (from the Implicit Function Theorem again)
that a local diffeomorphism is precisely a map between
manifolds of the same dimension which has maximal rank, or,
equivalently, that the pull-back of a local coordinate system
is a local coordinate system, which is the property we have
described for things π restricted to M'.

$\qquad$ To prove the Corollary to Theorem 4.1, let

$\qquad$ π': M' → B

be the map: π restricted to M'. We have seen that π is a
local diffeomorphism. Let U be an open subset of M, and

$\qquad$ φ: π(U) → U

a map which is the inverse of π restricted to φ. By its
definition, φ is the map described in the corollary.

Remark: It is obviously important to know when π': M' $\to$ B
is a global diffeomorphism. By Theorem 21.2 of DGCV a
sufficient condition for this is that:

    a)   B is simply connected, i.e. B is connected, and
        any closed curve can be deformed to a point.

    b)   There are a pair of points $(p_0, b_0) \in$ M' $\times$ B
        such that each curve in B beginning at B can be
        lifted under π' to a curve in M'.

If $B = R^n = M'$ (which is often satisfied in practice), b)
will be satisfied if

$$\lim_{x \to \infty} \pi'(x) = \infty.$$

It is a natural question (for economics certainly) to
inquire about which relative extrema of f are actually
minima. Here is one useful result.

Theorem 4.2. Keep the notations and hypotheses of Theorem
4.1. Suppose that $p_0 \in$ M' is a strong relative minimum of
$f_{\pi(p_0)}$. Let M_0' be the connected component of M' which
contains the point p_0. Then,

$f_{\pi(p)}$ has a strong relative

minimum at each point of M_0'.

Proof. Let $t \to p(t)$ be a curve, $0 \le t \le 1$, in M_0' such that:

$$p(0) = p_0.$$

We must show that $f_{(p(t))}$ has a strong relative minimum at $p(t)$, for all t. Here is the picture which goes along with this:

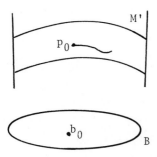

We must show that the Hessian of f is positive definite in a neighborhood, if it is positive definite at a point. This will follow from the following.

Exercise. Let $t \to A(t)$ $0 \le t \le 1$ be a symmetric, n × n real matrix, whose elements depend continuously on the parameter t. Suppose also that:

$$A(0)$$

is positive definite, and

$$\det (A(t)) \neq 0 \text{ for } 0 \leq t \leq 1.$$

Then, show that

$$A(t) \text{ is positive definite for all } t.$$

There is a generalization to other types of non-degenerate extrema.

Theorem 4.3.   With the hypotheses of Theorem 4.1, suppose that

$$p_0 \in M'$$

is a point such that the Morse index of $f_{\pi(p_0)}$ at p_0 is the integer r. Let M_0' be the connected component of M' containing p_0. Then, the Morse index of $f_{\pi(p)}$ at p, for all $p \in M_0'$, is precisely r.

Exercise.   Prove Theorem 4.3.

Of course, Theorem 4.2 is a special case of Theorem 4.3, with r = 0.

5. THE ECONOMICS OPTIMIZATION AS A CONTROL PROBLEM

Let us look at the material of Section 4 from the economics point of view, described very well (in classical language) by Samuelson [1]. Return to his notation, i.e. suppose given a real-valued function

$$f(x, \alpha), \tag{5.1}$$

with x are the variables one wishes to choose in some optimal way (for example "prices" one wishes to choose so as to optimize "profits").

This may be compared with the standard control theory problem. One is given variables (x) and (u), called <u>state</u> and <u>control</u> variables, and a system of ordinary differential equations:

$$\frac{dx}{dt} = h(x, u, t). \tag{5.2}$$

One wishes to choose a curve $t \to u(t)$ in the control space, to optimize some <u>performance criterion</u>, e.g.

$$\int L(x(t), u(t), t)dt, \tag{5.3}$$

where $t \to x(t)$ is the solution of (5.2), with the given u(t) and the initial state vector

$$x_0 \equiv x(0)$$

also given.

So far, the analogy between control theory and economics is not particularly close mathematically, beyond the fact that both use the same word, "optimization." However, they become closer when the concept of "feedback" is added to the control problem. (In engineering terminology, this connects on "open loop" problem to a "closed loop" one).

A feedback law for the control problem is a mapping

$$(x, t) \rightarrow \varphi(x, t) \equiv u(x, t)$$

of the state x time space into the control space. Such a feedback law is an optimal law if the solutions of the differential equations:

$$\frac{dx}{dt} = h(x, u(x(t), t), t) \tag{5.4}$$

are optimal curves for the performance criterion (5.3). (Precisely, I mean the following statement: Let $t \rightarrow \hat{x}(t)$ be a solution of the ordinary differential equations (5.4). Let $\hat{u}(t)$ be defined as:

$$\hat{u}(t) = \varphi(\hat{x}(t), t). \tag{5.5}$$

Then, $t \rightarrow (\hat{x}(t), \hat{u}(t))$ is an optimal curve for (5.3), in the usual calculus of variations sense.)

Remark: This can usefully be considered geometrically from the fiber space point of view. (x, u, t) are coordinates of a manifold M, (x, t) coordinates of a manifold B, the map (x, u, t) → (x, t) considered as a fiber space projection map M → B. <u>Feedback laws</u> are then cross-sections B → M. This geometrical approach has briefly been described in my paper given at the 1973 NATO Conference at Imperial College, Algebraic and Geometric Methods in Control Theory, Proceedings edited by D. Mayne, published by Reidel. The genesis of these ideas is clearly in the ideas of classical mechanics and the calculus of variations centering around the notions of "extremal field" and "Hamilton-Jacobi equation." In fact, these ideas have already been exploited for "practical" purposes (but not necessarily in the most satisfying geometric way) by R. Bellman, and his notion of "Dynamic Programming."

Here is obviously where the basic economic optimization problem, symbolized by (5.1), comes closer to the control theory. Let M be the manifold whose variables are (x, α), B the manifold whose variables are (α). Let $\pi: M \to B$ be the map $(x, \alpha) \to \alpha$. f then is interpreted as a map M → R. In Section 4 we have developed conditions that there exist a <u>cross-section map</u>

$$\varphi: B \to M$$

such that:

For $b \in B$, $\varphi(b)$ is an optimal point of
the function f_b, $\equiv f$ restricted to $\pi^{-1}(b)$.

This mapping obviously plays the same role in economics as
the optimal feedback laws play in control theory problems.
In view of this economics meaning, it might be called an
optimizing strategy.

This geometric view leads to many possibilities of
geometrization. For example, one might be interested in
approximate optimizing strategies, which would be a sequence

$$\varphi_1, \; \varphi_2, \cdots$$

of cross-section maps: $B \to M$ such that:

$$\lim_{j \to \infty} \varphi_j(b) = \varphi(b)$$

is an exact optimizing strategy. Much of the discipline
called numerical analysis may be regarded as the study of
such approximation sequences. For example, see the book by
Daniel, "The Approximate Minimization of Functionals,"
published by Prentice-Hall. The influence of Computer
technology is strongly felt here - for example one is
obviously interested in what might be vaguely thought of as
the "efficiency" of such approximations.

6. AN EXAMPLE FROM ECONOMICS - SAMUELSON'S TAX-RATE PROBLEM

Let us do now what Samuelson calls [1], p. 14 an Illus-
trative Tax Problem.

Let M denote the real interval

$$0 < x < \infty.$$

The coordinate x on M stands for the output of a firm.
Suppose given two functions:

$$C: M \rightarrow R$$

$$p: M \rightarrow R$$

called the cost function and production function. Let t be
a positive real variable, called the tax rate. (It plays
the role of a control variable in systems theory problems).

Now, for fixed t, define a function

$$\pi_t: M \rightarrow R,$$

called the revenue function, by the following formula:

$$\pi_t(x) = xp(x) - C(x) - tx. \tag{6.1}$$

Then,

$$\frac{\partial}{\partial x}(\pi_t(x)) = p + x\frac{dp}{dx} - \frac{dC}{dx} - t \tag{6.2}$$

$$\frac{\partial^2}{\partial x^2}(\pi_t(x)) = 2\frac{dp}{dx} + x\frac{d^2p}{dx^2} - \frac{d^2C}{dx^2}$$

Set:

$$f(x, t) = p(x) + x \frac{dp}{dx} - \frac{dC}{dx} - t \qquad (6.4)$$

$$= \text{right hand side of (6.2)}.$$

Let T denote the space of variables t, $0 < t < \infty$.

Definition. A <u>production policy</u> is a mapping

$$\varphi: T \to M \qquad (6.5)$$

such that:

$$f(\varphi(t), t) = 0 \qquad (6.6)$$

for all $t \to T$

φ is a locally <u>optimal production policy</u> if:

$$\frac{\partial f}{\partial x} (\varphi(t)) < 0 \qquad (6.7)$$

for all $t \in T$.

Remarks: Here is the "economic" motivation for these definitions. For a given value t of the "tax rate," <u>if</u> (6.7) and
(6.8) are satisfied, then the function

$$x \to \pi_t(x)$$

has a relative maximum at $x = \varphi(t)$. This is presumably how
a firm would choose its output, i.e. to maximize "revenue."

Here is another way of looking at this:

Let

M' = M × T = space of variables (x, t)   (6.8)

f is a map M' → R.   Set:

N = {(x, t) ∈ M': f(x, t) = 0}.

(6.6) says that the graph of the map φ: T → X lies on the
subset N.   Condition (6.7) is _implied_ by saying that:

$\frac{\partial f}{\partial x}$ as a function: M' → R, is _positive_ on N.

This problem can also be considered as a _contrained_
optimization problem, whose theory will be considered in
the next chapter.   Suppose that x, p, C, t are independent
variables, parameterizing, say, R.   Let

π: R^4 → R

be defined as:

π = xp - C - tx.

Now, optimize π, _constrained_ by the relation

p - p(x) = 0

C - C(x) = 0.

This approach is very analogous to the geometric approach
to Thermodynamics presented in GPS and volume IX.   I will
go into this approach in more detail later on.

Now, the question of most interest for economics in this
model is the _direction_ and _rate of change_ of production under

changes in the tax rate, <u>assuming</u> the production is chosen
so that revenue is optimal. Mathematically, this means that
we should calculate

$$\frac{\partial \varphi}{\partial t},$$

with particular attention to its sign. We can obviously do
this using by differentiating (6.6) with respect to t.

$$\frac{\partial f}{\partial x} (\varphi(t), t) \frac{d\varphi}{dt} + \frac{\partial f}{\partial t} = 0,$$

or

$$\frac{d\varphi}{dt} = - \frac{\partial f}{\partial t} (\varphi(t), t) / \frac{\partial f}{\partial x} (\varphi(t), t). \qquad (6.9)$$

Using (4.4),

$$\frac{\partial f}{\partial t} = - 1,$$

here (6.9) takes the following simple form:

$$\frac{d\varphi}{dt} = (\frac{\partial f}{\partial x} (\varphi(t), t))^{-1}, \qquad (6.10)$$

which is negative using (6.7). Hence:

<u>The optimal production decreases</u>
                                                    (6.11)
<u>as the tax rate is increased.</u>

As Samuelson emphasizes, the most important economics
moral from this example is the qualitative conclusion (6.11),
which follows only from the assumption that the production

function is chosen to be (locally) optimal, without any
special assumption about the specific form of C(x) and p(x).
Such a conclusion would obviously play the same role in
economics as the concept of "physical law" plays in physics.
For example, "conservation of energy" is a conclusion from
a specific mathematical model of physics (e.g. Hamilton's
equations).

Exercise.   Investigate the global properties of these
extremals.

Chapter 14

OPTIMIZATION OF FUNCTIONS UNDER CONSTRAINTS

1. INTRODUCTION

The general mathematical topic to be treated here -
optimization under constraints - obviously goes to the
heart of whatever mathematical content there is to economics.
Again, see Samuelson's book [1] for the extensive discussion
of the connection between this mathematical topic and ideas
of economics. In this chapter I shall emphasize the differential geometric aspects of the theory, restricting attention
to equality constraints. The case of inequality constraints -
which the economists sometimes call Kuhn-Tucker theory -
will be considered later.

Here is the general problem. Let M be a topological
space, and let

$$f, f_1, \ldots, f_n: M \to R \tag{1.1}$$

be real valued functions on M. Consider the problem of
optimizing

$$f \ ,$$

subject to the constraints

$$f_1 = 0 = \ldots = f_n. \qquad (1.2)$$

This means that we consider the subset

$$M \supset N = \{p \in M: f_1(p) = 0 = \ldots = f_n(p)\},$$

and restrict f to M, to obtain a function

$$f_N : N \to R$$

The extrema of f subject to constraints (1.2) are then the extrema of f_N on N. For example, a point p' $\in$ N is a relative minimum of f, subject to the constraints, if there is an open subset U $\subset$ M such that:

$$f(p') \leq f(p) \qquad (1.3)$$

for all p $\in$ U $\cap$ N .

We are interested in the case where M is a manifold, and the functions (1.1) are C^∞ (or sufficiently differentiable usually C^2 will do). As is well known, the classical technique to handle this problem is to introduce Lagrange multiplier variables $\lambda_1, \ldots, \lambda_n$, form the function

$$f + \lambda_1 f_1 + \ldots + \lambda_n f_n$$

on M $\times$ R^n, and find its extrema. I will assume that the reader is familiar with the motivation for introducing these multipliers, and will investigate their mathematical properties. In particular, we shall see that the λ's and f's should be interpreted as functions on the normal vector

bundle to N. There are many interesting geometric properties of this construction, and these properties often have
a basic importance for physical and economic applications.

2. THE LAGRANGE MULTIPLIER RULE ON MANIFOLDS

Let M be a manifold, and let $f, f_1, \ldots, f_n$ be real-
valued functions

$$M \rightarrow R.$$

Introduce new real variables

$$\lambda = (\lambda_1, \ldots, \lambda_n), \in R^n,$$

called Lagrange multipliers. Set:

$$E = M \times R^n.$$

Regard the λ's as real valued functions on R^n, hence also
as functions on E. Set:

$$f_E = f + \lambda_a f_a \in F(E)). \tag{2.1}$$

(Choose indices $1 \leq a, b \leq n$, and the summation convention
on these indices).

We shall investigate the extreme points of the function
f_E.

$$df_E = df + f_a d\lambda_a + \lambda_a df_a.$$

Thus, $df_E = 0$ at a point of E if and only if:

$$df + \lambda_a df_a = 0, \tag{2.2}$$

and

$$f_a = 0 \tag{2.3}$$

at that point. Let

$$\pi: E \to M$$

be the Cartesian projection map of $E \equiv M \times R^n$ onto M.

$$(p, \lambda) \to (p).$$

Set:

$$N = \{p \in M: f_a(p) = 0\} \tag{2.4}$$

$$f_N = \text{restriction of } f \text{ to } N. \tag{2.5}$$

Theorem 2.1. Suppose that N is a submanifold of M. Then, the projection under π of an extreme point of f_E is an extreme point of f_N.

Proof. $f_a = 0$ when restricted to N, hence also $df_a = 0$ when restricted to N. Suppose $(p, \lambda) \in E$ is an extreme point of f_E. (2.3) shows that

$$p \in N.$$

(2.2) shows that $df_N(p) = 0$, i.e. p is extreme point of f_N.

Remarks: There is a more classical way of looking at this result. Let (p, λ) be an extreme point of f_E. (p, λ) is then a result of optimizing an unconstrained function, namely

$$f_E.$$

The projection

$$\pi(p, \lambda) = p$$

is then an extremal of f subject to the constraints $\{f_a = 0\}$. We have converted the problem of finding a constrained extremum to the problem of finding an unconstrained one on a bigger space, i.e. the space obtained by adjoining the "Lagrange multipliers" (λ_a). This is the classical Lagrange multiplier rule. In essence, what has been done is to prolong the constrained optimization problem to one on a space setting over the original one. The advantages of the prolonged system is that it is unconstrained, which is pur- chased at the expense of adding more variables to optimize. "Prolongation" is a general geometric idea which is diffi- cult to make precise, but which plays a prominent role in the work of E. Cartan. Alternately, an algebraist might say that π defines a homomorphism between the optimization problem on E and that on M. In fact the notion of "homo- morphism" and "prolongation" are closely related, of course.

Here is a more traditional situation where one can see the relation. Let G, G' be groups. Of course a homomorphism is a map

$$G \to G'$$

which preserves the group multiplication. Suppose G acts as a transformation group on a space S_1 and G' acts as a transformation group on a space S'. A homomorphism from (G, S) to (G', S') is a pair of maps $G \to G'$, $S \to S'$, such that:

G → G' is a homomorphism in the algebraic ssnse.
The following diagram of maps is commutative:

$$G \times S \to S$$
$$\downarrow \quad \downarrow \quad \downarrow$$
$$G' \times S' \to S',$$

where the vertical arrows denote the given maps, and the horizontal arrows denote the maps defined by the transformation group action.

In this situation, the notion of "homomorphism" and "prolongation" essentially coincide.

Once interpreting the Lagrange multiplier rule in this quasi-algebraic way, the next step is to ask for conditions that it be onto, and also one-one. Here are the traditional answers.

<u>Theorem 2.2.</u> Let f, $f_1, \ldots, f_n$ be real-valued functions on a manifold M. Let N be a submanifold such that:

$$f_1 = 0 = \ldots = f_n \text{ on M.}$$

Let f_N = f restricted to N

$$E = M \times R^n$$

$$= \{(p, \lambda_1, \ldots, \lambda_n): p \in M, \lambda_1, \ldots, \lambda_n \in R\}$$

$$f_E = f + \lambda_1 f_1 + \ldots + \lambda_n f_n$$

$$\pi: E \to M \text{ the projection map}$$

$$(p, \lambda) \to p.$$

Suppose that the following condition is satisfied:

For each $p \in N$,

$$N_p = \{v \in M_p: df_1(v) = 0 = \ldots = df_n(v)\}. \qquad (2.6)$$

Then, π maps the set of extremal points of f_E <u>onto</u> the set of extremal points of f_N.

<u>Remark</u>: Recall the convention that N_p denotes the tangent space to the manifold N <u>considered</u> <u>as</u> a linear <u>subspace</u> of M_p, the tangent space to M. Thus, if $\alpha: N \to M$ is the map which <u>defines</u> M' as a submanifold - strictly speaking a submanifold <u>is</u> the map α - then the subspace if

$$\alpha_*(N_p) \subset M_{\alpha(p)}.$$

It is, of course, convenient for notational and intuitive purposes to identify $\alpha(p)$ and p, N_p and $\alpha_*(M_p')$. (Of course the very <u>condition</u> that α be a submanifold map is that α_* be one-one.)

   <u>Proof of Theorem 2.2</u>. $p \in N$ is an extremal point of f' if and only if

$$df(N_p) = 0. \tag{2.7}$$

   <u>Proof of 2.7</u>. Let $\alpha: N \to M$ be the explicit submanifold map. f_N is then $\alpha^*(f)$. $df_N(p)$ then equals $d(\alpha^*(f))(p)$, which is $\alpha^*(df)(p)$, hence

$$0 = \alpha^*(df)(p)(N_p) \tag{2.8}$$

is the extremal condition.

   By the definition of α^*, the right hand side of (2.8) is just

$$df(p)(\alpha_*(M_p')),$$

which leads, using the notational convention that $\alpha_*(N_p) \equiv N_p$, to (2.7).

   Suppose that $p \in N_p$ does satisfy (2.7). Without loss in generality, we can suppose that $df_1,\ldots, df_n$ are linearly independent at p. (Otherwise, just throw away the dependent ones). Let

$$df_1, \ldots, df_n, \theta_{n+1}, \ldots, \theta_m$$

be elements of M_p^d which form a <u>basis</u> of M_p^d. (Here we use the "trivial" - but extremely useful - fact that a set of linearly independent elements of a vector space can be expanded to give a basis). Then, $df(p)$ can be written as follows:

$$df(p) = \lambda_1^{\,o} df_1(p) + \ldots + \lambda_n^{\,o} df_n(p)$$
$$+ \lambda_{n+1}^{\,o} \theta_{n+1} + \ldots + \lambda_m^{\,o} \theta_m, \tag{2.9}$$

with real coefficients λ^o. (2.7) implies that:

$$(\lambda_{n+1}^{\,o} \theta_{n+1} + \ldots + \lambda_m^{\,o} \theta_m)(N_p) = 0. \tag{2.10}$$

Let

$$M_p'' = \{v \in M_p : \theta_{n+1}(v) = 0 = \ldots = \theta_m(v)\}. \tag{2.11}$$

The fact that $df_1(p), \ldots, \theta_m$ form a <u>basis</u> of M_p implies that:

$$M_p = N_p \oplus M_p''. \tag{2.12}$$

(2.10)-(2.12) imply that

$$\lambda_{n+1}^{\,o} \theta_{n+1} + \ldots + \lambda_m o \; \theta_m = 0.$$

$\theta_{n+1}, \ldots, \theta_m$ linearly independent imply that:

$$\lambda_{n+1}^{\,o} = 0 = \ldots = \lambda_m^{\,o}. \tag{2.13}$$

(2.13) implies that df can be written as:

$$df(p) = \lambda_1^{\,o} df_1 + \ldots + \lambda_n^{\,o} df_n. \tag{2.14}$$

(2.14), together with the condition that $p \in N$, i.e.
$f_1(p) = 0 = \ldots = f_n(p)$, implies that

$$(p, \lambda^0)$$

is now an extreme point of

$$f_E = f + \lambda_1 f_1 + \ldots + f_n,$$

i.e. $\pi: E \to M$ maps the extreme points of f_E <u>onto</u> those of f_N.

<u>Remarks and exercises</u>. There is a more elegant algebraic
way of proving this result. Set:

$$M_p'' = M_p/N_p .$$

Geometrically, M_p'' is the <u>normal vector space</u> to M' at the
point p.

Our hypothesis means that $df_1, \ldots, df_n$ pass to the quotient
to define linear forms on M_p'', denoted by $(df_1)'', \ldots, (df_n)''$.

<u>Exercise</u>. Show that condition (2.6) of our hypothesis is
equivalent to the condition that $(df_1)'', \ldots, (df_n)''$ <u>span</u>
$(M_p'')^d$ i.e. every element of the dual space $(M_p'')^d$ can be

written (not necessarily uniquely), as a linear combination
of them.

Exercise. Show that the real number $(\lambda_1^0, \ldots, \lambda_n^0)$ (Lagrange
multipliers, of course) which provide the extremal (p, λ^0)
to f_E are those such that:

$$(df)'' = \lambda_1^0 (df_1)'' + \ldots + \lambda_n^0 (df_n)^0.$$

Using the result of this exercise, we see an immediate
necessary and sufficient condition for the uniqueness of
the Lagrange multipliers is that $df_1, \ldots, df_n$ be linearly
independent at p. This leads to the following definitive
(and classical) result.

Theorem 2.3. Keep the hypotheses of Theorem 2.2. Suppose
in addition that following condition is satisfied:

> For each $p \in N$, $df_1(p), \ldots, df_n(p)$
> are linearly independent elements of M_p^d.   (2.15)

Then, $\pi: E \to M$ defines a one-one map from the extreme
points of f_E onto the extreme points of f

$$f_N \equiv f \text{ restricted to } N.$$

Exercise. Prove Theorem 2.3.

Here is another version, which is useful in practice.

Theorem 2.4. Let $f_1, \ldots, f_n$ be real valued, C^∞ functions
on a manifold M. Let N be the set of points of M on which
these functions are all zero. Suppose that:

$$df_1(p), \ldots, df_n(p)$$

are linearly independent elements of $M_p^{\,d}$ <u>at each point of M'</u>.
Then,

a) N can be made into a submanifold of M.

b) The hypothesis of Theorem 2.3 are satisfied, i.e.
$\pi: E \to M$ provides an "isomorphism" between the
extreme points of f_E and $f_N \equiv f$ restricted to N.
In other words, the Lagrange multiplier rule
"works," and gives <u>unique</u> Lagrange multipliers,
applied to the problem of optimizing f subject to
the constraints $f_1 = 0 = \ldots = f_m$.

Proof. a) follows from the Implicit Function Theorem.
(In fact, a standard argument shows that N can be made into
a <u>regularly embedded submanifold</u> of M, i.e. the topology on
the manifold M' is <u>identical</u> with the subspace topology,
i.e. a subset $O' \subset M'$ is open if and only if it equals
$O \cap M'$, where O is an open subset of M).

Remark: For the convenience of readers who are more comfortable in classical language, here is how these results
would look in case:

$$M = R^m = \text{space of variables } (x_i),$$
$$1 \leq i, j \leq m.$$

Choose a, $1 \leq a, b \leq n$.

$$N = \{x \in R^m: f_a(x) = 0\}.$$

$$f_E(x, \lambda) = f(x) + \lambda_a f_a(x).$$

Condition (2.15) is then as follows:

> For each $x \in N$, the rank of the
> $n \times m$ matrix $(\frac{\partial f_a}{\partial x_i}(x))$ is equal   (2.16)
> to n, i.e. is "maximal."

The equations for the extremals of f_E are then:

$$f_a(x) = 0 \tag{2.17}$$

$$\frac{\partial f}{\partial x_i} + \lambda_a \frac{\partial f_a}{\partial x_i} = 0. \tag{2.18}$$

These are (n + m) equations for (n + m) unknowns, (x, λ).

Exercise. Show that condition (2.16) is the condition that
the equations (2.17)-(2.18) be a regular system of (n + m)-
equations for (n + m)-unknown, i.e. have a non-singular
Jacobian matrix.

3.   THE ALGEBRAIC STRUCTURE OF A REAL QUADRATIC FORM

The simplest interesting situation involving optimization under equality constraints occurs where the function
$x \to f(x)$ is quadratic.   (In economics this is called quadratic programming.   Optimization of linear functions under
inequality constraints is called linear programming.)   In
this section I shall present some algebraic material about
real quadratic forms which will be needed for this study.
(The material to be presented here has already been presented
in scattered form in LMP, vol. II, and in volume II of this
series, but it will be convenient to gather together all the
facts we need.)

Let V be a real finite dimensional vector space.   Let

$$\beta: V \times V \to R$$

be an R-bilinear, symmetric map.   The real valued function

$$v \to \beta(v, v)$$

is called the quadratic form associated with β.   This form
uniquely determines β, and it will be convenient for us to
identify the bilinear and quadratic forms.

For the rest of this system let such a form β be fixed.

Let V^d denote the dual space of V, the vector space of
linear maps

$$V \to R.$$

For $v \in V$, let

$$v \lrcorner \ \beta \in V^d$$

be the element defined as follows:

$$(v \lrcorner \beta)(v_1) = \beta(v, v_1) \qquad (3.1)$$

for all $v_1 \in V$.

Set:

$$\ker (\beta) = \{v \in V: v \lrcorner \beta = 0\}. \qquad (3.2)$$

Definition. β is said to be <u>non-degenerate</u> (or <u>non-singular</u>)
if:

$$\ker (\beta) = 0. \qquad (3.3)$$

Exercise. Show that β is non-degenerate if and only if the
linear map

$$v \to v \lrcorner \beta$$

of $V \to V^d$ is an isomorphism.

Definition. Let V_1 be a linear subspace of V. The <u>orthogonal
complement</u> of V_1, denoted by $V_1^{\perp}$, is defined as:

$$V_1^{\perp} = \{v \in V: \beta(v, V_1) = 0\}. \qquad (3.4)$$

Theorem 3.1. If V_1 is a linear subspace of V, then

$$\dim (V_1^{\perp}) \geq \dim V - \dim V_1. \qquad (3.5)$$

Proof. Let $v_1, \ldots, v_m$ be a basis of V_1. Consider:

$$\theta_1 = v_1 \lrcorner \beta \in V^d$$

$$\vdots$$

$$\theta_n = v_n \lrcorner \beta \in V^d$$

(3.4) means that $V_1^{\perp}$ is the set of $v \in V$ such that:

$$\theta_1(v) = 0 = \ldots = \theta_n(v). \qquad (3.6)$$

Suppose: $\dim V = m$. Then, (3.6) represents n linear homo-geneous equations in m unknowns.

Exercise. Show that the dimension of the space of solutions of a system of linear equations like (3.6), involving n equations in m unknown, is at least m-n.

The result of this exercise now proves (3.5).

Theorem 3.2. Let V_1 be a linear subspace of V. Then, β restricted to V_1 is non-degenerate if and only if

$$V = V_1 \oplus V_1^{\perp}, \qquad (3.7)$$

i.e. V is a direct sum of the subspaces V_1, $V_1^{\perp}$.

Proof. Suppose first that β, restricted to V is non-degenerate. Then,

$$V_1 \cap V_1^{\perp} = (0). \tag{3.8}$$

(For, if (3.8) were not satisfied, i.e. a non-zero $v \in V_1$ and $v \in V_1^{\perp}$, then

$$\beta(v, V_1) = 0,$$

i.e. v would be in the kernel of β restricted to V_1).

Consider the map

$$V_1 \times V_1^{\perp} \to V$$

which assigns to $(v_1, v_2) \in V_1 \times V_1'$ their sum $v_1 + v_2$. This map is a _linear_ map of the direct sum of the vector spaces V_1, $V_1^{\perp}$. Condition (3.8) means that the kernel of the map is zero, i.e. it is one-one. Relation (3.5) implies that:

$$\dim (V_1 \times V_1^{\perp}) \geq \dim V.$$

From linear algebra we know that a one-one linear map of one vector space to another, with the _dimension of the domain greater or equal to the dimension of the range_, is an _isomorphism_.

Exercise. Prove this fact about linear maps between _finite dimensional_ vector spaces.

This observation now proves (3.7).

Now, let us prove the converse. Suppose (3.7) is satisfied. Then β restricted to V_1 must be non-degenerate - otherwise, $V_1 \cap V_1^{\perp}$ would be non-zero.

Remark: Theorem 3.2 is really the Fundamental Theorem in the theory of algebraic structure of the "structure" of quadratic forms. It also holds for quadratic forms with arbitrary fields of scalars, and for skew-symmetric bilinear forms. See Artin's book "Geometric algebra" [1], for a systematic exposition of this "structure" theory.

Now we turn to study properties of quadratic forms which really involve the properties of the real numbers. Consider the case:

$$V = R^m = \{x = (x_1, \ldots, x_m); x_1, \ldots, x_m \in R\}.$$

Then, β takes the following more traditional form:

$$\beta(x, x) = \sum_{1 \leq i, j \leq m} \beta_{ij} x_i x_j. \tag{3.9}$$

(β_{ij}) is a symmetric, $m \times m$ matrix, called the matrix of the quadratic form. The properties of the quadratic form can be systematically studied in terms of matrix theory. (See Gantmacher's treatise on Matrix Theory). However, the coordinate-free techniques of modern linear algebra are much more efficient, as we shall see.

It was well-known in classical, 19th century algebra,

how to deal with (3.9) - reduce it to a "canonical form."
The 19th century algebraists (Cayley, Sylvester,...) showed
that a new <u>linear</u> coordinates could be chosen for R^m so that:

$$\beta(x, x) = x_1^2 + \ldots + x_r^2 - x_{r+1}^2 - \ldots - x_{r+r}^2. \quad (3.10)$$

This is the <u>canonical form</u>. The integer s is usually called
the <u>index of inertia</u>. <u>Sylvester's Theorem</u> asserts that the
s and r are <u>invariants</u> of the quadratic form, i.e. if β is
reduced to such a sum and difference of squares in two
different ways, the integers r and s are the same in both
cases. In matrix language this means that the matrix can
be reduced to <u>diagonal</u> form

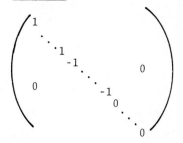

We shall now deal with these facts using the theory of vector
spaces. Return to a bilinear form $\beta: V \times V \to R$.

<u>Definition</u>. β is <u>positive definite</u> if:

$$\beta(v, v) > 0 \qquad\qquad\qquad (3.11)$$

for all $v \in V-(0)$.

β is <u>negative definite</u> if - β is positive definite.

A positive definite form is of course non-degenerate. (Because, if $v \in ker(\beta)$, then $\beta(v, v) = 0$, which forces $v = 0$). In terms of the canonical form (3.10)

$$s = 0, \; r = m$$

in the positive definite case. In Volume II, Chapter 4, I have proved that the necessary and sufficient condition for positive definiteness of β is the positivity of the <u>principal minor determinants</u> of the matrix (β_{ij}), i.e.

$$\beta_{11} > 0$$

$$\begin{vmatrix} \beta_{11} & \beta_{12} \\ \beta_{21} & \beta_{22} \end{vmatrix} > 0. \qquad\qquad (3.12)$$
$$\vdots$$

<u>Definition</u>. A linear subspace V_- is a <u>maximal negative subspace</u> of V if the following conditions are satisfied:

a)   β is negative definite on V_-

b) V_- is contained in no larger linear subspace with property a).

<u>Exercise</u>. Show that such maximal negative subspaces <u>exist</u>.

(Hint: Choose V_- as a subspace of <u>maximal dimension</u> among the set of negative subspaces).

<u>Theorem 3.3</u>. Let V_- be a maximal negative subspace of V. Let

$$V_-^\perp$$

be its orthogonal complement with respect to β. Then,

$$V = V_- \oplus V_-^\perp \qquad\qquad (3.13)$$

β is positive semi-definite
on $V_-^\perp$, i.e.   (3.14)
$\beta(v, v) \geq 0$ for $v \in V_-^\perp$.

The kernel of β restricted to
$V_-^\perp$ is equal to the kernel of β itself.   (3.15)

   <u>Proof</u>. The direct sum decomposition (3.11) follows from Theorem 3.2, since β is non-degenerate when restricted to V_-.

   Suppose (3.14) is <u>not</u> satisfied, i.e. there exists a $v \in V_-^\perp$ such that

$$\beta(v, v) < 0.$$

Then, $V_-^\perp + (v)$ would be a linear subspace on which β would be negative definite, contradicting the maximality of V_-.

Exercise.   Prove (3.15), to complete the proof of Theorem
3.3.

Definition.   A triple (V_+, V_-, V_0) of linear subspaces of V
defines a canonical decomposition of V if the following conditions are satisfied:

β is positive definite on V_+ (3.16)

β is negative definite on V_- (3.17)

V_0 is the kernel of β (3.18)

$\beta(V_+, V_-) = 0.$ (3.19)

Theorem 3.4.   Such canonical decompositions exist.   If
(V_+, V_-, V_0), (V_+', V_-', V_0) are two such decompositions,
then

$\dim V_+ = \dim V_+'$ (3.20)

$\dim V_- = \dim V_-'$ (3.21)

V_- is a maximal negative subspace   (3.22)

V_+ is a maximal positive subspace   (3.23)

Remark: (3.20) and (3.21) are just the vector space version
of Sylvester's classical Theorem of Inertia.   We shall call

the integer

$$s = \dim V_-, \tag{3.24}$$

which is an invariant of β, the _index_ of the quadratic form
β. We shall see that it plays a basic role in optimization
theory. It is equal to the integer s which appears in the
classical canonical form (3.10).

Proof of Theorem 3.4. First, we prove existence of a
canonical decomposition. Let V_- be any maximal negative
subspace. By Theorem 3.3,

$$\ker(\beta) \equiv V_0 \subset V_-^{\perp},$$

and β is positive semidefinite on $V_-^{\perp}$. Let V_+ be any maximal
positive subspace of β restricted to $V_-^{\perp}$.

Exercise. Show that:

$$V_-^{\perp} = V_+ \oplus V_0. \tag{3.25}$$

Show also that V_+ is a maximal positive subspace of V.

(3.25) now proves the _existence_ of a canonical decomposition. Suppose $(V_+, V_-, V_0, (V_+', V_-', V_0)$ are two
canonical decompositions. Let

$$\pi_-: V \to V_-$$

$$\pi_+: V \to V_+$$

$$\pi_0 \colon V \to V_0$$

be the projection maps in this decomposition, i.e.

$$1 = \pi_0 + \pi_+ + \pi_-, \tag{3.26}$$

$$(\pi_+)^2 = \pi_+, \ (\pi_-)^2 = \pi_-, \ (\pi_0)^2 = \pi_0.$$

To prove. π_- restricted to V_-' is one-one.

Suppose $v \in V_-'$ is such that

$$\pi_-(v) = 0.$$

In view of (3.26), this means that:

$$v \in V_+ + V_0.$$

Hence, $\beta(v, v) \geq 0$. But, $v \in V_-'$ means that $\beta(v, v) \leq 0$, which forces

$$\beta(v, v) = 0,$$

which forces $v = 0$, which implies (3.26).

We deduce from (3.26):

$$\dim (V_-') \leq \dim V_-. \tag{3.27}$$

Reversing the role of the primed and unprimed canonical decompositions proves the reverse inequality to (3.27), hence proves (3.21).

Exercise.   Prove (3.22) and (3.23), to complete the proof of Theorem 3.4.

Exercise. If V_- is any maximal negative subspace, show that
it is a part of at least one canonical decomposition.

Exercise. Derive the classical canonical form (3.8) from
Theorem 3.4.

Exercise. If (V_+, V_-, V_0), (V_+', V_-', V_0) are two canonical
decompositions, show that there is a linear map

$$\varphi: V \to V$$

such that:

$$\varphi(V_+) = V_+', \quad \varphi(V_-') = V_-'$$

$$\varphi(V_0) = V_0$$

φ is an isomorphism of β, i.e.

$$\beta(\varphi(v_1), \varphi(v_2)) = \beta(v_1, v_2)$$

$$\text{for all } v_1, v_2 \in V.$$

Remark: This is a special (easier) case of Witt's Theorem.
See LMP, vol. II, Chapter III, Section 19, and Lang's
treatise on Algebra [1].

Maximal Isotropic subspaces

This is a topic which will be important later, when we
study the Lagrange multiplier aspects of minimization problems.

It will give us an algebraic tool to prove the Kuhn-Tucker Theorem in a simple but important special case.

Continue with β as a quadratic form on the finite dimensional vector space V. Suppose β is non-degenerate. Then, $V_0 = (0)$, and the canonical decomposition takes the form:

$$V = V_+ \oplus V_-, \tag{3.28}$$

$$(V_+, V_-) = 0 \tag{3.29}$$

$s = \dim (V_-)$ is called the index of β.

Remark: The group of automorphisms of β is then isomorphic to the matrix group

$$O(m-s, s), \text{ with } m = \dim V,$$

using the notations explained in LMP, vol. II. For example,

$$O(3, 0) \equiv O(3, R)$$

is the 3-variable orthogonal group, $O(1, 3)$ is the Lorentz group, and so forth.

Definition. A linear subspace $I \subset V$ is said to be isotropic if:

$$\beta(I, I) = 0. \tag{3.30}$$

Such a subspace is said to be maximal isotropic if it is

contained in no larger isotropic subspace.

<u>Theorem 3.5.</u> Suppose that (3.26) is satisfied, with:

$$s = \dim(V_-) \leq r = \dim(V_+). \qquad (3.31)$$

Then, the dimension of a maximal isotropic subspace of V is equal to s.

<u>Proof.</u> Again, let

$$\pi_+ : V \to V_+$$

$$\pi_- : V \to V_-$$

be the projection maps of V with respect to the direct sum decomposition (3.26).

<u>To prove</u>: π_+ and π_- restricted to I are one-one. (3.31)
Suppose otherwise, e.g. $v \in I$,

$$\pi_-(v) = 0. \quad \text{Then,}$$

$$v \in V_+, \text{ here } \beta(v, v) > 0,$$

which contradicts that $v \in I$, i.e. contradicts (3.30). A similiar argument holds if $\pi_+(v) = 0$.

(3.31) - plus the fact that a one-one linear map does not decrease dimension - implies that:

$$\dim I \leq s.$$

If dim I < s, then

$$\pi_-(I) \neq V_-.$$

Let $v_- \in V_-$ be such that:

$$\beta(v_-; \pi_-(I)) = 0, \quad \beta(v_-, v_-) = -1.$$

Similarly, $\pi_+(I) \neq V_+$. Choose $v_+ \in V_+$ such that:

$$\beta(v_+, \pi_+(I)) = 0, \quad \beta(v_+, v_+) = 1.$$

Then,

$$I + (v_+ + v_-)$$

is an isotropic subspace which contains I, i.e., I is not maximal.   q.e.d.

Remark: If (3.27) is not assumed, then here is the more general way of stating the result:

$$\dim I = \min (\dim (V_+), \dim (V_-)). \tag{3.32}$$

The sign of a quadratic form

This topic will also be important in optimization. It is already discussed in Volume II, Chapter 4, but I shall go over it again.

Let β be a quadratic form on a vector space V. Let

$$(v_i), \quad 1 \leq i, j \leq m \; = \dim M$$

be an basis of V. Then, the

(m × m) real matrix

$$\beta_{ij} \equiv \beta(v_i, v_j) \tag{3.33}$$

is called the <u>matrix</u> of β with respect to this basis.

Now, the determinant

$$\det(\beta_{ij})$$

of this basis is <u>not</u> independent of the choice of bases, as
is the determinant of a linear map. However, if the basis
is changed

$$v_i' = \alpha_{ij} v_j,$$

the determinant changes as follows

$$\det(\beta_{ij}') = \det(\alpha_{ij}))2 \ \det(\beta_{ij}) \tag{3.34}$$

In particular, the <u>sign</u> of the determinant of the quadratic
form is invariant.

<u>Definition</u>. The <u>sign</u> of β is a number which is 0, ±1,
defined as follows:

0 if β is degenerate, i.e. $\det(\beta_{ij}) = 0$

1 if $\det(\beta_{ij}) > 0$

-1 if $\det(\beta_{ij}) < 0$.

Theorem 3.6. Suppose $V = V_1 \oplus V_2$, with $\beta(V_1, V_2) = 0$. Let $\beta_1 = \beta$ restricted to V_1, $\beta_2 = \beta$ restricted to V_2. Then,

$$\text{sign of } \beta = (\text{sign of } \beta_1)(\text{sign of } \beta_2). \qquad (3.35)$$

Exercise. Prove (3.35). It follows from the usual properties of determinants, particularly the rule:

$$\det \begin{pmatrix} A_1 & 0 \\ 0 & A_2 \end{pmatrix} = \det(A_1) \det(A_2).$$

Theorem 3.7. Suppose β is non-degenerate, and has index s. Then,

$$\text{sign } \beta = (-1)^s. \qquad (3.36)$$

Exercise. Prove (3.36).

Exercise. Let β be the quadratic form on R^3 whose matrix is

$$\begin{pmatrix} 1 & 2 & -3 \\ 2 & 5 & -4 \\ -3 & -4 & 8 \end{pmatrix}.$$

Compute a canonical decomposition of R^3, and compute the maximal isotropic spaces.

Let us now turn to some applications of this algebraic formalism.

4. CONDITIONS THAT A CONSTRAINED QUADRATIC FORM BE POSITIVE DEFINITE

Now we shall discuss an algebraic problem which is important in the theory of constrained extrema. (See Samuelson [1], p. 376). It will be presented first purely algebraically then its relation to constrained extrema problems will be considered more explicitly.

Let V be a real vector space, and let

$$\beta: V \times V \to R$$

be a symmetric, bilinear form on V.

Now, suppose that (θ_a), $1 \leq a \leq m$, are elements of V^d, i.e. are linear maps

$$\theta_a: V \to R.$$

We suppose that the θ_a are linearly independent elements of V^d.

Question. What are the conditions, that β be positive definite on the linear subspace of the $v \in V$ such that

$$\theta_a(v) = 0?$$

To answer this, let us make a "Lagrange multiplier" construction. Set:

$$E = V \times R^n. \tag{4.1}$$

Denote a point of E by

$$(v, \lambda),$$

with $v \in V$, $\lambda = (\lambda_1, \ldots, \lambda_n) \in R^n$. Define the function

$$f_E: E \to R$$

as follows:

$$f_E(v, \lambda) = \beta(v, v) + \lambda_a \theta_a(v). \qquad (4.2)$$

If f_E a quadratic form on the <u>vector space</u> E? Yes.
Define

$$\beta_E: E \times E \to R$$

as follows:

$$\begin{aligned} \beta_E((v, \lambda), (v', \lambda')) &= \beta(v, v') \\ &+ \frac{1}{2} \lambda_a \theta_a(v') \qquad (4.3) \\ &+ \frac{1}{2} \lambda'_a \theta_a(v). \end{aligned}$$

Then,

$$f_E(v, \lambda) = \beta_E(v, \lambda), (v, \lambda)), \qquad (4.4)$$

which show that f_E is indeed the quadratic form associated
with the bilinear form β_E.
Set:

$$V' = \{v \in V: \theta_1(v) = 0 = \ldots = \theta_n(v)\}. \qquad (4.5)$$

Theorem 4.1. β_E is non-degenerate if and only if β restricted to V' is non-degenerate.

Proof. Suppose that

$$(v, \lambda) \in \text{kernel } \beta_E.$$

In view of (4.3), this means that:

$$\beta(v, v') + \frac{1}{2} \lambda_a \theta_a(v')$$
$$+ \frac{1}{2} \lambda_a' \theta_a(v) = 0 \qquad (4.6)$$

for all $(v', \lambda') \in E$.

The arbitrary nature of (λ_a') first forces:

$$\theta_a(v) = 0, \qquad (4.7)$$

i.e.

$$v \in V'.$$

With condition (4.7) holding, (4.6) now implies that

$$v \in \text{kernel}(\beta \text{ restricted to } V'). \qquad (4.8)$$

Thus, (β restricted to V non-degenerate) implies β_E non-degenerate.

Conversely, suppose that $v \in V'$ satisfies (4.8). Since $\theta_a = 0$ defines V', and the θ_a are linearly independent, there are real numbers λ_a such that

$$v \lrcorner \beta = -\lambda_a \theta_a. \qquad (4.9)$$

Exercise, Prove the existence of λ_a. This implies (4.6), i.e. $(v, \lambda) \in$ kernel β_E.   q.e.d.

Remark: Notice a new facet of the marvelous "Lagrange multiplier" technique. Given a form β and a linear subspace $V' \subset V$, we have "prolonged" β to be a form β_E on a vector space E, which "sits over V", in the sense that there is a projection map

$$\pi: E \to V,$$

This assignment

$$(\beta, V') \to \beta_E \qquad\qquad (4.10)$$

has the property that β_E is non-degenerate if and only if β restricted to V' is non-degenerate. In fact, the assignment (4.10) is a typical example of a "functor."

Exercise. Make more explicit the category-functor aspect of the Lagrange multiplier construction.

Suppose now that β_E is non-degenerate. We shall find the conditions that it must satisfy in order that β restricted to V' be positive or negative definite.

First, identify V with the subspace of E consisting of the pairs of the form

$$V = \{(v, 0): v \in V\}.$$

β_E restricted to this subspace is obviously β.

Also, set:

$$\Lambda = \{(0, \lambda): \lambda \in R^n\}.$$

E is then a direct sum of these linear spaces.

$$E = V \oplus \Lambda.$$

We see from (4.3) that:

$$\beta_E(\Lambda, \Lambda) = 0,$$

i.e.

<u>Λ is an isotropic subspace of E.</u>   (4.11)

Here is the basic result.

<u>Theorem 4.2</u>. Λ is a maximal isotropic subspace of E if
and only if β restricted to V' is positive or negative
definite.

<u>Proof</u>. We have just seen that Λ is isotropic. Suppose
it is not maximal isotropic. There is then an element
$(v, \lambda) \in E$ such that

$$(v, \lambda) \notin \Lambda, \text{ i.e. } v \neq 0,$$

but $\Lambda + (v, \lambda)$ is also isotropic. This means that

$$\beta_E((v, \lambda),(0, \lambda')) = 0 \qquad (4.12)$$

for all $\lambda' \in R^n$, and

$$0 = \beta_E((v, \lambda), (v, \lambda)). \tag{4.13}$$

Using (4.3), (4.12) now translates into the following condition:

$$\lambda_a{}' \theta_a(v) = 0, \tag{4.14}$$

Since (4.14) holds for all $\lambda' \in R^n$,

$$\theta_a(v) = 0, \text{ or}$$

$$v \in V'. \tag{4.15}$$

Condition (4.13) now means that:

$$\beta(v, v) + \lambda_a \theta_a(v) = 0. \tag{4.16}$$

Combining (4.15) and (4.16), leads to the condition

$$\beta(v, v) = 0. \tag{4.16}$$

Suppose now that β restricted to V' is positive or negative definite. (4.16) implies $v = 0$, i.e. Λ is maximal isotropic. Conversely, suppose that β restricted to V' is not positive or negative. As we have seen in Section 3, there then exists a non-zero $v \in V'$ which satisfies (4.16). The steps leading to (4.16) are readily seen to be reversible, i.e. there is a λ such that $\Lambda + (v,\lambda)$ is isotropic.

All the basic properties of the Lagrange multiplier construction now follow from Theorem 4.2, plus the standard

facts about the algebraic structure of quadratic forms described in Section 3. For example:

<u>Theorem 4.3</u>. β restricted to V' is positive definite if and
only if the index of β_E is equal to n.

   <u>Proof</u>. Suppose that:

$$\dim V = m.$$

Then,

$$\dim V' = m - n. \tag{4.17}$$

(To prove (4.17), note that V' is determined by the n independent linear equations $\theta_a = 0$). Suppose first that:
<u>β restricted to V' is positive definite</u>.
By Theorem 4.2, we know that n is equal to the dimension of
a maximal isotropic subspace of E. By the results of
Section 3, we can find a decomposition:

$$E = E_+ \oplus E_-, \tag{4.18}$$

such that β_E is positive definite on E_+, negative definite
on E_-, and

$$\beta_E(E_+, E_-) = 0. \tag{4.19}$$

We can also suppose that

$$V \subset E_+. \tag{4.20}$$

(To prove (4.20), extend V - identified with the subspace
$\{(v, 0)\}$ of E - to a maximal subspace E_+ on which β_E is
positive definite, and let E_- be the orthogonal complement
of E_+).

By definition of the "index," we have:

$$s \equiv \text{index } (\beta_E) = \dim E_-.$$

Also, as proved in Section 3,

$$n = \text{dimension of maximal isotropic subspace}$$
$$= \text{minimum of } (\dim E_+, \dim E_-). \tag{4.21}$$

We also know (from (4.4) again) that:

$$\beta_E(V', \Lambda) = 0. \tag{4.22}$$

As in Section 3, let:

$$\pi_+: E \to E_+,$$

$$\pi_-: E \to E_-$$

be the projection maps of E with respect to the direct sum
decomposition (4.18).

We have seen in Section 3 that π_+ and π_- are one-one
when restricted to Λ. Now, $V' \subset E_+$. Hence, (4.22) implies
that $\pi_+(\Lambda)$ is perpendicular to V'.

In particular, $\pi_+: \Lambda \to E_+$ cannot be onto. But, we have seen
in Section 3 that in order that Λ be maximal isotropic
either π_+ of π_- must be onto. We can conclude that

$$\pi_- : \Lambda \rightarrow E_-$$

is onto, i.e. π_- is an isomorphism between Λ and E_-. In particular,

$$\dim \Lambda = n = \dim E_- \equiv s = \text{index of } \beta_E.$$

This concludes the proof of Theorem 4.3 in one direction.

Exercise. Provide the converse argument, to conclude the proof of Theorem 4.3.

Remarks. a) Algebraic significance

Theorem 4.3 may be regarded as the Main Theorem from the applications point of view. (Of course, Theorem 4.2 is the Main Theorem from the mathematical structure view point, and   Theorem 4.3 is a easy consequence of it). Theorem 4.3 gives us a way of calculating whether β restricted to V' is positive definite - compute the index of β_E and see whether or not it equals n. Here is a qualitative result which enhances the interest in this criterion - roughly, it says that the criterion given is stable.

Exercise. If β_E of index n, and is non-degenerate, show that all quadratic forms on E which are sufficiently close to β_E in the natural topology for quadratic forms (e.g. so

that their matrix coefficients are close when a basis is
chosen for E) are also of index n. (More generally, the
non-degenerate real quadratic forms of a given index form
an open subset in the space of all real quadratic forms).

b)   Minimization under constraints and the Kuhn-
      Tucker Theorem

These results have a great importance in optimalization
theory. To see this directly, suppose

$$f: V \to R$$

is the real-valued function such that

$$f(v) = \beta(v, v),$$

i.e. f is the quadratic form defined by β.

Exercise. Show that:

$v = 0$ is a minimum of f restricted to the subspace V' (or constrained by the linear equation,
$\theta_a(v) = 0$) if and only if β is positive definite
when restricted to V'.

Consider

$$f_E: E \to R,$$

defined by

$$f_E((v, \lambda)) = \beta_E((v, \lambda), (v, \lambda))$$

for $(v, \lambda) \in E$.

Then, β_E has index n if and only if f_E has a saddle point at $(0, 0) \in E$, i.e. in m directions going out from 0 f_E increases, while in n direction f_E decreases.

Exercise. Make this precise in terms of Morse Theory, i.e. show that f_E has a non-degenerate extreme point of Morse index n if and only if f has a non-degenerate minimum at $v = 0$. This result is a typical example of the Kuhn-Tucker Theorem [1], which plays a key role in the mathematics of economics.

c) The intrinsic form of the Lagrange-Multiplier
    construction-extensions to infinite dimensions

We can now clarify the algebraic nature of the construction. This has a great advantage for the purpose of generalization - for example to manifolds, or to infinite dimensional linear spaces.

Let V be a vector space, not necessarily finite dimensional. Let V' be a linear subspace. Let $\beta: V \times V \to R$ be a symmetric bilinear form. Let Λ be the space of linear maps

$$\lambda: V \to R$$

such that:

$$\lambda(V') = 0.$$

(In case the quotient space if finite dimensional, Λ is just the dual space of V/V', i.e. $(V/V')^d$. In infinite dimensional situations, one must take care to impose topological conditions correctly.) Set:

$$E = V \times \Lambda \equiv \text{"abstract" direct sum of}$$
$$\text{the vector spaces } V \text{ and } \Lambda.$$

(Of course, V and Λ may then be identified with the appropriate linear subspaces of E). Define

$$\beta_E: E \times E \to R$$

as follows:

$$\beta_E((v_1, \lambda_1), (v_2, \lambda_2)) = \beta(v_1, v_2)$$
$$+ \frac{1}{2} \lambda_1(v_2) + \frac{1}{2} \lambda_2(v_1)$$
$$\text{for } v_1, v_2 \in V; \lambda_1, \lambda_2 \in \Lambda.$$

Notice now that everything we have done carries over.

Exercise. In case V is finite dimensional, prove the results of this Section in this basis - independent setup. Investigate possible generalizations to infinite dimensional situations. In case V is finite dimensional, with V' defined by $\theta_a = 0$, note that $\lambda \in \Lambda$ is identified with $\lambda_a \theta_a$, i.e. the Lagrange multiplier parameters $(\lambda_1, \ldots, \lambda_n)$ are basically just linear functions on V/V'.

d)   The classical form of the conditions

Suppose now that

$$V = R^m = \{(x_1, \ldots, x_m): x_1, \ldots, x_m \in R\}.$$

Choose indices and summation conventions as follows:

$$1 \le i, j \le m; \ 1 \le a \le n.$$

Suppose:

$$\beta(x, x) = \beta_{ij} x_i x_j,$$

with $\beta_{ij} = \beta_{ji}$.

$$\theta_a = \alpha_{ai} x_i.$$

Then, E is the space of variables (x, λ). β_E is the form:

$$(x, \lambda) \to \beta_{ij} x_i x_j + \alpha_{ai} \lambda_a x_i.$$

The $(n + m) \times (n + m)$ symmetric matrix of β_E then takes the following block form:

$$\begin{pmatrix} \beta, & \alpha^T \\ \alpha, & 0 \end{pmatrix}, \tag{4.23}$$

where β is the m × m matrix (β_{ij}), α the n × m matrix (α_{ai}), α^T its transpose. In Samuelson's book [1], "Foundations of Economic Analysis," p. 378, one will find the condition that β restricted to V' be positive definite expressed in terms of the signs of certain subdeterminants of this matrix.

$$0 < (-1)^n \begin{vmatrix} \beta_{ij} & \alpha_{ai} \\ \alpha_{ia} & 0 \end{vmatrix}, \tag{4.24}$$

$$1 \le i, j \le r, \ 1 \le a, b \le n$$

$$r + 1 \le r \le m.$$

The derivation of these equations is of algebraic interest, and will now be considered.

5. THE PRINCIPAL MINOR POSITIVITY CONDITIONS FOR QUADRATIC FORMS WITH CONSTRAINTS

Because of its importance for the applications, e.g. to economics, I will now go into detail about the algebraic ideas underlying the derivation of the positivity conditions (4.24).

Let us recapitulate the construction, using the "intrinsic" form of the definition. V is a finite dimensional real vector space; V' is a linear subspace;

$$\beta : V \times V \to V$$

is a symmetric bilinear form on V.

$$E = V \times (V/V')^d.$$

β_E is the form on E defined by

$$\beta_E((v_1, \lambda_1)(v_2, \lambda_2)) \tag{5.1}$$

$$= \beta(v_1, v_2) + \frac{1}{2}\lambda_1(v_2) + \frac{1}{2}\lambda_2(v_1)$$

for $v_1, v_2 \in V$, $\lambda_1, \lambda_2 \in (V/V')^d$.

We have seen in Section 4 that β restricted to V' is positive definite if and only if the index of β_E is equal to n, where:

$$n = (\dim V - \dim V'). \tag{5.2}$$

Recall (from Section 3) that the <u>sign</u> of β_E is the sign (i.e. ± 1) of the determinant of β_E with respect to <u>any</u> bases of E.

Theorem 5.1. If β restricted to V' is positive definite, then:

$$\text{sign } (\beta_E) = (-1)^n. \tag{5.3}$$

Proof. We have seen in Section 4 that β_E is non-degenerate, and of index n. This means that there is a basis of E such that the matrix of β_E in this basis takes the diagonal form:

$$\begin{pmatrix} 1 & & & & & \\ & \ddots & & & & \\ & & 1 & & & \\ & & & -1 & & \\ & & & & \ddots & \\ & & & & & -1 \end{pmatrix} \tag{5.4}$$

with n minus signs. This of course implies (5.3)

Now, let us suppose that

$$V_1 \subset V$$

is a linear subspace of V of <u>codimension one</u>, i.e.

$$1 = \dim V - \dim V_1. \tag{5.5}$$

Set:

$$V_1' = V_1 \cap V'. \tag{5.6}$$

Suppose that:

$$V_1' \text{ is of codimension one in } V'. \tag{5.7}$$

<u>Remark</u>: If V_1 is in "general position" with respect to V, then (5.7) will be satisfied. Note that (5.7) also means that:

$$\dim (V/V') = \dim (V_1/V_1'). \tag{5.8}$$

This can take a more explicit form in terms of linear equations. Suppose

$$\theta_1, \ldots, \theta_n$$

are linearly independent linear forms on V whose vanishing characterizes V', i.e. V' is defined by n linear equations in m (= dim V) unknown. Then, the "general position" condition means that $\theta_1, \ldots, \theta_n$ <u>remain</u> linearly independent when restricted to V_1.

Now, set:

$$\beta^1 = \beta \text{ restricted to } V_1. \tag{5.9}$$

$$E_1 = V_1 + (V/V_1')^d. \tag{5.10}$$

Recall (from Section 4) that $(V/V')^d$ is identified with the space of linear maps

$$\lambda: V \to R$$

such that:

$$\lambda(V') = 0.$$

We can restrict such a λ to V_1, obtaining a linear map: $V_1 \to R$ which is zero on $V_1 \cap V' \equiv V_1'$. This assignment then defines a linear map

$$(V/V')^d \to (V_1/V_1')^d. \tag{5.11}$$

Exercise. Show that (under the "general position" condition (5.8)) the map (5.11) is an isomorphism.

Assuming (5.11) is a isomorphism, we can then identify

$$E_1 = V_1 \times (V_1/V_1')^d$$

with a linear subspace of

$$E = V \times (V/V')^d.$$

Theorem 5.2. With the bilinear form $\beta_{E_1}^{\ 1}$ defined on E_1 by

the formula analogous to (5.1), β_E restricted to the subspace E_1 is equal to $\beta_{E_1}^{\;\;1}$.

The proof is implicit in the remarks made above ex-
plaining how E_1 is a subspace of E.

We then have the following result, which is the Main
Result of this section.

Theorem 5.3. If β restricted to V' is positive definite,
and if V_1 is a codimension one linear subspace in general
position, with respect to V', then:

$$\text{sign } (\beta_E) = {}^{\prime} \; (-1)^n = \text{sign } \beta_{E_1}^{\;\;1}. \qquad (5.12)$$

Proof. If β is positive definite on V', then β^1 is
positive definite on

$$V_1' = V_1 \cap V'.$$

But we know that the condition for this is that the index of
$\beta_{E_1}^{\;\;1}$ be equal to $(-1)^{(\dim V_1 - \dim V_1')}$, which is equal to
$(-1)^n$ by (5.8).

Now we can iterate Theorem 5.3. Choose a "flag" linear
subspace

$$V \supset V_1 \supset V_2 \supset \ldots \supset V_{m-n-1} \qquad (5.13)$$

each of which is of codimension one in the preceding one,
and each of which is in general position with respect to V'.

Set:

$$\beta^2 = \beta \text{ restricted to } V_2$$
$$\beta^3 = \beta \text{ restricted to } V_3 \tag{5.14}$$
and so forth.

$$\beta_{E_2}^{\ 2} = \text{form constructed using Lagrange}$$
$$\text{multiplier formula (5.1) and so forth.}$$

<u>Theorem 5.4.</u> β restricted to V' is positive definite <u>if</u> <u>and only if</u>

$$(-1)^n = \text{sign } (\beta_E) = \text{sign } \beta_{E_1}^{\ 1} = \ldots \tag{5.15}$$

$\underline{\text{Proof.}}$ If β is positive definite, then (5.15) results from iterating (5.15). To see that (5.15) is <u>sufficient</u> for positivity of β restricted to V', note that:

$$\dim V_1' = m - n - 1$$
$$\dim V_2' = m - n - 2$$
$$\vdots \tag{5.16}$$
$$\dim V_{m-n-1}' = 1.$$

Hence,

$$0 \subset V_{m-n-1}' \subset \ldots \subset V'$$

provides a "flag" of linear subspaces of V', starting from 0 and ending up with V'. (5.15) implies that β is positive

definite on V'_{m-n-1}, since it is of dimension one. Applied
again, it implies that β restricted to V_{m-n-2} is positive
definite. Continuing in this way, we reach V_1'. Notice
that this is a generalization of the argument given in
Volume II, Chapter 4, to cover the positivity of a form on
a vector space with no constraints, which is the special case:

$$V = V'.$$

Exercise. Referring back to the details given in Volume II,
Chapter 4 (using the multiplicative property of the sign of
a quadratic form on the direct sum of two vector spaces)
give the details of this argument.

Exercise. Prove the classical formula (4.23) has specializing
Theorem 4.4. (Hint: $V = R^n$, $= \{(x_1, \ldots, x_n)\}$,

$$V_1 = \{x \in R^n : x_1 = 0\}$$
$$V_2 = \{x \in R^n : x_1 = 0 = x_2\}$$

and so forth.

Exercise. Work out the conditions in case β is defined by
the matrix

$$\begin{pmatrix} 1 & -3 & 2 \\ -3 & 7 & -5 \\ 2 & -5 & 8 \end{pmatrix},$$

and V' is defined by the constraint

$$x_1 + x_2 = 0.$$

We can now return to study the Hessian of a general function on a manifold subject to constraints.

6. SECOND ORDER CONDITIONS FOR CONSTRAINED EXTREMALS AND DUALITY

Since all of our work is local, it is just as easy to work in the "classical" situation, with functions on R^n, rather than the general manifold case. Denote a point of R^n by

$$x = (x_1, \ldots, x_n) \equiv (x_i),$$

$$1 \leq i, j \leq n.$$

Let

$$f: R^n \to R$$

be a real valued function.

Adopt indices as follows:

$$1 \leq a, b \leq m.$$

Suppose given functions

$$g_a: R^n \to R,$$

the <u>constraint functions</u>.

<u>Problem</u>.   <u>Extremize f, under the constraints $g_a = 0$</u>.

We have seen how the Lagrange multiplier rule works. Introduce new variables:

$$y = (y_a) \in R^m.$$

Define a function

$$F(x, y) = f(x) + y_a g_a. \tag{6.1}$$

F maps $R^n \times R^m \to R$.

The extremal points of f constrained by $g_a = 0$ are the projections into R^n of the extremal points of F in R^n.

Suppose

$$(x^o, y^o) \in R^n \times R^m$$

is an extremal point of F.

Let us compute the Hessian of F at the extreme point (x^o, y^o), a routine calculation using (6.1).

$$\frac{\partial^2 F}{\partial x_i \partial x} = \frac{\partial^2 f}{\partial x_i \partial x_j} + y_a \frac{\partial^2 g_a}{\partial x_i \partial x_j}$$

$$\frac{\partial^2 F}{\partial y_a \partial x_i} = \frac{\partial g_a}{\partial x_i}$$

$$\frac{\partial^2 F}{\partial y_a \partial y_b} = 0.$$

Introduce $v = (v_i)$ as coordinates of the tangent vectors to R^n, $\lambda = (\lambda_a)$ as coordinates to the tangent vectors to R^m. Then,

$$(v, \lambda)$$

are coordinates of the tangent vectors of $R^n \times R^m$.

Remark: In other words, a tangent vector to $R^n \times R^m$ is identified with the differential operator

$$v_i \frac{\partial}{\partial x_i} + \lambda_a \frac{\partial}{\partial y_a}.$$

In terms of these linear coordinates of tangent vectors to $R^n \times R^m$, we can write down the Hessian to F at the extremal point (x^o, y^o).

Theorem 6.1. Let $(x^o, y^o) \in R^n \times R^m$ be an extremal point of the function

$$(x, y) \to F(x, y) = f(x) + y_a g_a,$$

i.e. a solution of the following equations:

$$g_a(x^o) = 0 \qquad\qquad\qquad (6.2)$$

$$\frac{\partial f}{\partial x_i}(x^o) + y_a^{\,o} \frac{\partial g_a}{\partial x_i}(x^o) = 0. \qquad\qquad (6.3)$$

Then, the Hessian form of F at (x^o, y^o) is given by the

following formula:

$$(v, \lambda) \to h(v, \lambda)$$

$$= (\frac{\partial^2 f}{\partial x_i \partial x_j} (x^0) + y_a^0 \frac{\partial^2 g_a}{\partial x_i \partial x_j} (x^0)) v_i v_j \qquad (6.4)$$

$$+ \frac{\partial g_a}{\partial x_i} (x^0) \lambda_a v_i.$$

We can interpret (6.4) algebraically. Construct a quadratic form Q: $R^n \to R$ and linear form θ_a: $R^n \to R$ as follows:

$$Q(v) = (\frac{\partial^2 f}{\partial x_i \partial x_j} (x^0) + y_a^0 \frac{\partial^2 g_a}{\partial x_i \partial x_j} (x^0)) v_i v_j \qquad (6.5)$$

$$\theta_a(v) = \frac{\partial g_a}{\partial x_i} v_i. \qquad (6.6)$$

Notice that (6.4) can be written as follows:

$$h(v, \lambda) = Q(v) + \lambda_a \theta_a(v). \qquad (6.7)$$

Here is the main result of this section:

Theorem 6.1. Set:

$$N = \{x \in R^n : g_a(x) = 0\}. \qquad (6.8)$$

Suppose that $dg^0(x^0)$ are linearly independent, so that N is a submanifold of R^n of dimension n-m in a neighborhood of x^0. Then, the quadratic form h on $R^n \times R^m$ has index m if and only if f restricted to n has a strong relative minimum at x^0.

Proof. Algebraically we know (after the work of Sections 3 - 4) that h given by formula (6.7) has index m if and only if the following condition is satisfied:

Q is positive definite when restricted
to the subspace of $v \in R^n$ such that

$$\frac{\partial g_a}{\partial x_i} (x^o)v_i = 0. \tag{6.9}$$

Notice that the linear subspace of R^n determined by condition (6.9) is essentially tangent space to N at x^o.

Since h is the Hessian of F at an extremal point, we know that it is _independent of coordinates_. Let us choose coordinates conveniently. Since we assume dg_a to be linearly independent, we know (implicit function theorem) that the coordinates for R^n can be chosen so that:

$$g_a = x_a.$$

Thus, the Hessian of f restricted to N is the quadratic form

$$\frac{\partial^2 f}{\partial x_i \partial x_j}, \quad m + 1 \leq 1, \; j \leq n.$$

Conditions (6.9) become

$$v_a = 0.$$

With this choice of g_a,

$$\frac{\partial^2 g_a}{\partial x_i \partial x_j} = 0,$$

hence Q reduces to the Hessian of f, namely:

$$v \rightarrow \frac{\partial^2 f}{\partial x_i \partial x_j} v_i v_j .$$

Theorem 6.1 now follows:

Corollaries to Theorem 6.1. Suppose that x^o is a strong
relative minimum of f restricted by the constraints $g_a = 0$.
Then, the function $F(x, y)$ has a saddle-point type of extremal
at (x^o, y^o). For n independent directions leading out of
x^o, y^o, F increases, and for m independent directions leading
out of (x^o, y^o) into $R^n \times R^m$ F decreases. In particular,
there is an m-dimensional submanifold M of $R^n \times R^m$ containing (x^o, y^o) such that:

> (x^o, y^o) is a strong maximum
>
> of F restricted to M.

Thus, (x^o, y^o) is a solution of a constrained maximum
problem, called the Lagrange dual optimization problem to
the one with which we started. (See Section 7 for a general
explanation of what is involved here.)

Remark: The Corollaries are basically a differential-
geometric version of the Kuhn-Tucker Theorem [1], which plays
such a important role in applications to economics.

7. SADDLE POINTS AND DUAL VARIATIONAL PROBLEMS

In section 6, we saw that a constrained minimization problem leads to an unconstrained saddle-point type of extremal in the corresponding Lagrange multiplier problems. In this chapter, I will return to general manifold - Morse theory ideas to explain what is involved here.

Let M be a manifold,

$$f: M \to R$$

a real-valued, C^∞ function on M. Recall that a point $p \in M$ is an <u>extremal point</u> of f if:

$$df(p) = 0, \qquad\qquad (7.1)$$

i.e. if $v(p) = 0$ for all $v \in M_p$.

<u>Remark</u>: It is more usual to call p a <u>critical point of f</u>. However, in the calculus of variations one speaks of "extremals," not "criticals," and I prefer a terminology which is uniform for the finite and infinite dimensional cases:

Supposing that $p \in M$ is such an extremal point, we can define a symmetric, bilinear form

$$h_f: M_p \times M_p \to R,$$

called the <u>Hessian</u> of f. It has the following two important properties:

$$h_f(X(p), Y(p)) = X(Y)(f))(p)$$

$$\text{(7.2)}$$

for $X, Y \in V(M)$.

If $t \to \sigma(t)$ is a curve in M, $-1 < t < 1$, such that:

$$\sigma(0) = p_i \sigma'(0) = v \in M_p, \text{ then:}$$

$$\frac{d^2}{dt^2} f(\sigma(t))/_{t=0} = h_f(v, v).$$

$$\text{(7.3)}$$

The extremal point p is said to be <u>non-degenerate</u> if the Hessian form h_f if non-degenerate. It follows that p is then an <u>isolated</u> extremal point. There are obviously other relations between the algebraic properties of h_f and the geometric properties of the extremal. For example:

If h_f if positive definite, then p is
a strong relative <u>minimum</u> of f

If h_f is negative definite, then p
is a strong relative maximum of f

Of course, if h_f if non-degenerate, but neither positive nor negative, then p as a <u>saddle-point</u> extremal, i.e. in same directions, going out from p, f increases, in others f decreases, just as one finds at a "saddle" or "col" between two mountains

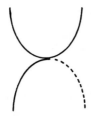

<u>Example</u>: $M = R^2 = \{(x, y): x, y \in R\}$.

$f(x, y) = x^2 - y^2$.

x - 0, y = 0 is a saddle point f increases along the curve

$x = t, \; g = 0,$

decreases along the curve

$x = 0, \; g = t.$

Let us make this more precise:

<u>Definition</u>. Let N be a submanifold of M passing through the non-degenerate extremal point p. N is said to be a <u>minimum</u> <u>submanifold</u> relative to f of the following conditions are satisfied:

$$h_f(v, v) > 0 \text{ for all}$$
$$v \in N_p. \tag{7.4}$$

N_p is contained in no larger linear
subspace of M_p on which h_f is positive
definite, i.e. on which (7.4) is   (7.5)
satisfied.

Similiarly, N is called a <u>maximal submanifold</u> for f at p if
N is a minimal manifold for -f.

Here are the main geometric properties:

<u>Theorem 7.1</u>. If N is a minimal submanifold of f, then p is
a strong relative minimum of f restricted to N. If N is a
maximal submanifold, then p is a strong relative maximum of
f restricted to N.

<u>Theorem 7.2</u>. There are a pair (N^+, N^-) of a minimal and a
maximal submanifold for f at p such that:

$$\dim M = \dim N^+ + \dim N^- \tag{7.6}$$

$$N_p^+ \cap N_p^- = (0). \tag{7.7}$$

The dimension N^- is equal to the <u>index</u> of h_f, i.e. the
maximal number of minus signs in a canonical form for h_f.

<u>Exercise</u>. Prove Theorems 7.1 - 7.2.

Remark: These properties suggest the typical "duality" between variational problems. For, we have:

$$f(p) = (\text{local}) \min_{q \in N^+} f(q) \qquad (7.8)$$

$$f(p) = (\text{local}) \max_{q \in N^-} f(q). \qquad (7.9)$$

In other words, by constraining f to N^+, we convert the problem of extremizing f to a minimization problem, and constraining to N^- defines the dual maximization problem. Of course, this split-up into dual max and min problems is not unique.

These ideas may be readily extended to infinite dimensional spaces, e.g. to calculus of variations problems. Te They lead directly to what are called complementary variational problems in Applied Mathematics. (See Arthurs [1]). Later on I shall describe this topic in terms of the geometric calculus of variations theory being developed in this work.

Chapter 15

PARETO OPTIMALS AND EXTREMALS

1. INTRODUCTION

Economists like to define their discipline as "the science of allocating scarce resources." A key problem is then the study of ways of allocating objects to "consumers" so that it is not possible to make some better off without making others worse off. Such an allocation is called a Pareto Optimal allocation.

As we have seen, the study of the optimization of a single function on a manifold (corresponding, say, to an economic problem of allocation to a single consumer) leads to a rich and interesting geometric structure. The Pareto problem is a natural generalization, but has been little studied by mathematicians. (Recent interesting work by Smale and other differential topologists is beginning to fill in this gap.) In this chapter I will develop some elementary, but basic, ideas.

2. THE FIRST ORDER PARETO CONDITIONS

Let M be a manifold, and

f: M → R

a C$^\infty$ function. When we study the optimization of f, the
"first order conditions" are the points x ε M such that

$$df(x) = 0 .$$ (2.1)

<u>Remark</u>. In order to avoid confusion with the notation "p"
for "prices", I start to denote points of manifolds with
other letters. Unfortunately, it is not possible to be
completely consistent in these notational matters!

It is a very familiar advanced-calculus story how the
conditions (2.1) are put together with "second order" conditions - thought of geometrically as "curvature" or "convexity"
conditions - to study true optima.

Now suppose

$$f_1,\ldots,f_n: M \to R$$

are a collection of real-valued functions.

Consider the following "economic" interpretation of
these functions:

A point x ε M represents the goods allocated to
n consumers, labelled 1,...,n .

f_1 measures the benefit to the first consumer.

f_2 measures the benefit to the second consumer.

and so forth.

Definition. A point $x \in M$ is a __Pareto optimum__ of the functions $(f_1, \ldots, f_m)$ if there do __not__ exist points $x' \in M$ such that:

$$f_1(x) < f_1(x'), \ldots, f_m(x) < f_m(x'). \qquad (2.2)$$

Definition. A point $x \in M$ is called a __Pareto extremum__ of $(f_1, \ldots, f_n)$ if the following condition is satisfied:

The subspace of M_x^d spanned by $df_1, \ldots, df_n$ at x is __at most__ $\qquad (2.3)$ $(n-1)$-dimensional.

Theorem 2.1. If a point $x \in M$ is a Pareto optimum of $(f_1, \ldots, f_n)$, then it is a Pareto extremum.

Proof. Suppose x is not a Pareto extremum, i.e. the $df_1, \ldots, df_n$ are linearly independent at x. Set:

$$M' = \{x' \in M: f_2(x) = f_2(x'), \ldots, f_n(x) = f_n(x')\} \qquad (2.4)$$

By the Implicit Function Theorem, M' is a submanifold of M, at least in a neighborhood of x.

Let:

$$f' = f_1 \text{ restricted to } M'. \qquad (2.5)$$

I assert that:

$$x \in M' \text{ must be an extremum of } f'. \qquad (2.6)$$

Otherwise, there are points $x' \in M'$ close to x such that:

$$f'(x') > f'(x). \qquad (2.7)$$

Combine (2.5) and (2.7):

$$f_1(x') > f_1(x).$$

By (2.4), $x' \in M$ means that:

$$
\begin{aligned}
f_2(x) &= f_2(x') \\
&\vdots \\
f_n(x) &= f_n(x').
\end{aligned}
\qquad (2.8)
$$

(2.7) and (2.8) contradict Pareto optimality of x, implying (2.6). But, (2.6) means that:

$$df' = 0 \text{ at } x',$$

i.e.

$$
\begin{aligned}
df_1 &= \text{linear combination of} \\
&\quad df_2, \ldots, df_n \text{ at } x,
\end{aligned}
\qquad (2.9)
$$

which contradicts our initial assumption that $df_1, \ldots, df_n$ were independent at x.

Remark: In economics terms, this argument means: Hold constant the allocations to the last $(n-1)$ consumers, and vary the first, thus reducing the problem to that of optimizing a single function.

Here is another necessary condition for pareto optimality:

Theorem 2.2. Suppose that $x \in M$ is a Pareto Optimum of $(f_1, \ldots, f_n)$. Let $(f_1, \ldots, f_r)$ be chosen (by reordering the functions, if necessary) so that:

$$df_1, \ldots, df_r \text{ are linearly} \tag{2.10}$$
$$\text{independent at } x$$

$$r = \text{dimension of } M_x^d \text{ spanned} \tag{2.11}$$
$$\text{by } df_1, \ldots, df_n.$$

Then, at x, there are relations of the following form:

$$df_{r+1} = a_{r+1,1} \, df_1 + \ldots + a_{r+1,r} \, df_r$$
$$\vdots \tag{2.12}$$
$$df_n = a_{n1} df_1 + \ldots + a_{nr} df_r.$$

Set:

$$\underset{\sim}{a}_1 = (a_{r+1,.}, \ldots, a_{n1}) \in R^{n-r}$$
$$\underset{\sim}{a}_2 = (a_{r+1,2}, \ldots, a_{n2}) \in R^{n-r} \tag{2.13}$$
$$\vdots$$
$$\underset{\sim}{a}_r = (a_{r+1,r}, \ldots, a_{nr}) \in R^{n-r}.$$

Then,

$$\underset{\sim}{a}_1 \not> 0, \ldots, \text{ and } \underset{\sim}{a}_r \not> 0. \tag{2.14}$$

Proof. The existence of (2.12), with real coefficients a, follows merely from linear algebra, and our assumptions (2.10) and (2.11).

Suppose that (2.14) is not satisfied. For example, suppose that:

$$a_1 > 0, \text{ i.e.}$$
$$a_{r+1,1} > 0, \ldots, a_{n1} > 0. \tag{2.15}$$

There is then a curve

$$t \to x(t)$$

in M starting at x, such that:

$$\frac{d}{dt} f_1(x(t))/_{t=0} > 0,$$
$$\frac{d}{dt} f_i(x(t)) = 0 \text{ for } 2 \leq i \leq r.$$

Using (2.12), we then have:

$$\frac{d}{dt} f_i(x(t))/_{t=0} > 0 \text{ for } r + 1 \leq i \leq n,$$

which contradicts Pareto optimality of x.

Corollary to Theorem 2.2. Suppose that x is Pareto optimal for $(f_1, \ldots, f_n)$, that the dimension of the subspace of M_x^d spanned by $df_1, \ldots, df_n$ is equal to (n-1). Then, for $a = (a_1, \ldots, a_n) \in R^n$ such that:

$$a_1 df_1 + \ldots + a_n df_n = 0 \text{ at } x,$$

$$\underset{\sim}{a} \geq 0 \text{ or } \underset{\sim}{a} \leq 0, \qquad\qquad (2.16)$$

i.e. the $a_1, \ldots, a_n$ all the same sign.

Remark: In interpreting (2.14) or (2.16), we adopt the notational convention (which is standard in the economics literature) that:

$$\underset{\sim}{v} \geq 0$$
$$\text{for } \underset{\sim}{v} = (v_1, \ldots, v_n) \in R^n$$

means that: $v_1 \geq 0, \ldots, v_n \geq 0$.

$$\underset{\sim}{v} > 0$$

means that $v_1 > 0, \ldots, v_n > 0$. This

$$\underset{\sim}{v} \not> 0$$

means that $\underset{\sim}{v}$ is not > 0, i.e. one of the components is non-positive.

There are various ways to rephrase these results so that they can be more readily remembered. To do so, introduce the following notations:

$$a = (a_1, \ldots, a_n) \quad R^n \qquad\qquad (2.17)$$

$$f = (f_1, \ldots, f_n) \qquad\qquad (2.18)$$

$$f \cdot a = a_1 f_1 + \ldots + a_n f_n. \qquad\qquad (2.19)$$

Exercise. Show that $x \in M$ is a Pareto extremal point of $\mathfrak{f}$ if and only if there is a non-zero vector $\mathfrak{a} \in R^n$ such that

$$\mathfrak{f} \cdot \mathfrak{a}$$

has an extremum at x in the ordinary sense, i.e.

$$d(\mathfrak{f} \cdot \mathfrak{a})(x) = 0. \qquad (2.20)$$

Show further that if x is a Pareto optimum of $\mathfrak{f}$, there is a non-zero vector $\mathfrak{a} \in R^n$ satisfying (2.20), and the following further condition:

$$\mathfrak{a} \geq 0. \qquad (2.21)$$

3. THE LOCAL PROPERTIES OF PARETO OPTIMUM POINTS

Continue with the notations of Section 3. The term Pareto optimum is abbreviated to

P.O.

Pareto extremum is abbreviated to

P.E.

R_+^n denotes the set of real n-vectors

$$\mathfrak{a} = (a_1, \ldots, a_n)$$

such that

$$\mathfrak{a} > 0,$$

i.e.

$$a_1 > 0, \ldots, a_n > 0$$

Definition. If x is a Pareto extremum point of functions
$\underset{\sim}{f} = (f_1, \ldots, f_n)$, define the rank of x as the dimension of
the subspace of M_x^d spanned by

$$df_1, \ldots, df_n,$$

i.e. the maximal number of linearly independent covectors
among this set. x is said to be a P.E. of maximum rank if
its rank is n-1.

Exercise. Show that the P.E. points of maximal rank form
an open subset of the set of P.E. points.

Exercise. If x is a P.O. point of maximal rank, show that
there is an

$$\underset{\sim}{a} \in R_+^n$$

such that

$$d(\underset{\sim}{f} \cdot \underset{\sim}{a})(x) = 0. \tag{3.1}$$

Theorem 3.1. If x is a P.O. point of maximal rank, and
$\underset{\sim}{a} \in R_+^n$ satisfies (3.1), then x is a maximum point of the
function $\underset{\sim}{f} \cdot \underset{\sim}{a}$.

Proof. Suppose otherwise, i.e. there is a point $x' \in M$ such that

$$\text{\bf f} \cdot \text{\bf a}(x) < \text{\bf f} \cdot \text{\bf a}(x').$$

Then, given the meaning, (2.19), of $\text{\bf f} \cdot \text{\bf a}$, we have:

$$a_1 f_1(x) + \ldots + a_n f_n(x) < a_1 f_1(x') + \ldots + a_n f_n(x'), \tag{3.2}$$

or

$$a_1(f_1(x) - f_1(x')) + \ldots + a_n(f_n(x) - f_n(x')) < 0. \tag{3.3}$$

Now, the $a_1, \ldots, a_n$ are all, by hypothesis, positive. Hence,

$$f_1(x) < f_1(x'), \ldots, f_n(x) < f_n(x'),$$

i.e. x is not a P.O. point.

Here is the converse.

Theorem 3.2. Suppose $\text{\bf a} \in R_+^n$ is such that $x \in M$ is a maximum point for $\text{\bf f} \cdot \text{\bf a}$. Then, x is a Pareto optimal point of $\text{\bf f}$.

Remark: The flavor of these results is that the study of Pareto optima is equivalent to the study of ordinary optima of families of functions, parameterized by convex subsets of R_+^n. It is then natural from a geometric point of view to introduce the Hessian of $\text{\bf f} \cdot \text{\bf a}$ as the second order "invariant" of P.E. points.

<u>Definition</u>. For x ∈ M, let

$$R^n(\underline{f}, x) = \{\underline{a} \in R^n : d(\underline{f} \cdot \underline{a})(x) = 0\}. \qquad (3.4)$$

$$R_+^{\,n}(\underline{f}, x) = R^n(\underline{f}, x) \cap R_+^{\,n}. \qquad (3.5)$$

Suppose x ∈ M is a P.E. point for $\underline{f}$, i.e. $R^n(\underline{f}, x) \neq 0$.
Then, the linear map

$$R^n(\underline{f}, x) \rightarrow M_x^{\,d} \circ M_x^{\,d}, \qquad (3.6)$$

defined as follows:

> The image under the map (3.6) of a
> point x ∈ R^n is the Hessian form
> at x of the function $\underline{f} \cdot \underline{a}$,

is called the <u>Hessian</u> of the P.E. point.

<u>Remark</u>: Recall that the Hessian form of a function f: M → R
that has an extreme point at x ∈ M is a symmetric bilinear
form

$$M_x \times M_x \rightarrow R.$$

As usual in multilinear algebra (see Volume II, or Graeb
[1]), we identify such a bilinear form with an element of
$M_x^{\,d} \circ M_x^{\,d}$, the <u>symmetric tensor product</u> of the covector
space $M_x^{\,d}$ with itself. In the language of tensor analysis,
the elements of $M_x^{\,d} \circ M_x^{\,d}$ are <u>symmetric, second-order, covariant tensors at x.</u>

Here is one typical way the Hessian may be used to study the Pareto optimality properties:

Theorem 3.3. Suppose x is a Pareto extremum of f of maximum rank. Suppose that the Hessian form (3.6) is negative definite, then x is a local Pareto optimum, i.e. there is an open neighborhood of x such that f restricted to this neighborhood is Pareto optimal.

Remark: The "negativeness" of the Hessian form (3.6) means that, the following two conditions are satisfied:

> $R_+^n(f, x) \neq 0$.
> For $a \in R_+^n(f, x)$ (which is determined up to a positive scalar multiple, since x is a maximal rank P.E.), the Hessian form of $f \cdot a$ is negative definite in the usual sense.

Exercise. Prove Theorem 3.3.

Here is a related idea:

Definition. A Pareto extremum point $x \in M$ is non-degenerate if the following conditions are satisfied:

> x is a maximal rank P.E.

For each $a \in R^n$ such that $d(f \cdot a)(x) = 0$,

the Hessian form of $f \cdot a$ at x is non-degenerate.

Here is another typical result.

Theorem 3.4. Let N be the subsets of point $x \in M$ which are
non-degenerate Pareto extreme points for functions

$$f = (f_1, \ldots, f_n).$$

Then, N is a submanifold of M, with

$$\dim N = n-1. \tag{3.7}$$

Remarks: Here is the intuitive reason for the formula (3.7)
for dimension of N. Given $(c_1, \ldots, c_{n-1}) \in R^n$, consider
the subset of $x \in M$ such that:

$$f_1(x) = c_1, \ldots, f_{n-1}(x) = c_n.$$

Call this subset:

$$M(c_1, \ldots, c_{n-1}).$$

On each such subset, pick out an extreme point of f_n restricted to this subset. Assume it is isolated. (This
corresponds to the "non-degeneracy" condition). As
$(c_1, \ldots, c_{n-1})$ vary, there extreme points sweep out the
Pareto extremum, i.e. define a local parameterization of N
by the (n-1) real parameters $(c_1, \ldots, c_{n-1})$. This is the

intuitive reason for (3.7).

Here is the picture in case

dim M = 2, n = 2

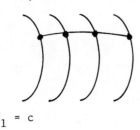

$f_1 = c$

The vertical curves are the subsets

$f_1 = \text{constant} = c_1$.

The dots $\cdot$ on these curves denote the optima of f_2 restricted
to these curves. As c_1 varies, these dots sweep out the
horizontal curve, which is the set of Pareto extrema.

Proof of Theorem 3.4.

This just involves putting in analytical form the
geometric idea described above.

Let x be a point of N. Suppose the ordering of $(f_1, \ldots, f_n$
is chosen so that:

$df_1, \ldots, df_{n-1}$ are linearly
independent at x.

We can also suppose - at most subtracting from f_n a linear combination of the $f, \ldots, f_{n-1}$, that:

$$df_n = 0 \text{ at } x.$$

In order to define a coordinate system for N, choose a coordinate system for M. Say,

$$(x_i), \ 1 \leq i, \ j \leq m = \dim M.$$

Suppose also that the coordinates of the point x are 0.

Introduce real parameters

$$\lambda = (\lambda_1, \ldots, \lambda_{n-1}) \in R^{n-1}.$$

Consider the following functions on $R^m \times R^{n-1}$:

$$F_i(x, \ \lambda) = \frac{\partial f_n}{\partial x_i} + \lambda_1 \frac{\partial f_1}{\partial x_i} + \ldots + \lambda_{n-1} \frac{\partial f_{n-1}}{\partial x_i} \qquad (3.8)$$

Now, our assumption that the point $x \in M$ is a non-degenerate P.E. means that:

$$\det\left(\frac{\partial f_n}{\partial x_i \partial x_j}(0)\right) \neq 0. \qquad (3.9)$$

Exercise. Show that (3.9) implies that the functions F_i are functionally independent.

Notice now that the relations

$$F_i(x, \ \lambda) = 0, \qquad (3.10)$$

for (x, λ) is a sufficiently small neighborhood of $(x, \lambda) = (0, 0)$, characterize points of N, i.e. non-degenerate

P.E. points, since at such a point

$$d(\underset{\sim}{f} \cdot \underset{\sim}{a}) = 0,$$

where:

$$\underset{\sim}{a} = (\lambda_1, \ldots, \lambda_{n-1}, 1).$$

We have shown that each point of N can be parameterized by functionally independent relations (3.10). This (using the implicit function theorem) gives a local parameterization of N by

$$m + (n-1) - m = n-1$$

real parameters.

Exercise. Furnish the proof that N is a submanifold of M. (In the sense that N can be given a manifold structure such that the inclusion map

$$N \to M$$

is one-one on tangent vectors.) Is this a "regular embedding", i.e. is the topology defined by the manifold structure on N the same as the topology induced on N as a subspace of M?

Remark: Let N' be the set of the non-degenerate Pareto optimum points. It appears that, in general at least, N' is diffeomorphic to R^{n-1}. Here is a sketch of how such a

proof might go (or alternately an intuitive reason for such
a conjecture).

Consider the set of

$$a \in R_+^{\,n}$$

such that:

$$a_1^{\,2} + \ldots + a_n^{\,2} = 1.$$

Exercise. Show that this subset of R^n is diffeomorphic to
R^{n-1}. Assign to each such a the optimal point $x(a) \in M$

$$f \cdot a$$

Supposed unique. We know that such a point is P.O. point,
i.e. a point of N'. One might expect that it would be possible to prove, at least in certain circumstances, that the map

$$a \to x(a)$$

is a diffeomorphism between N' and R^{n-1}. For example, if
the following conditions are satisfied, I believe such a
proof would go through. (Exercise)

$$M = R^m$$

Each function $f_1, \ldots, f_n$ are strongly concave,
i.e. their Hessian matrix is negative definite.

Such a condition seems typical of situations in economics.

Final Remarks

(March 1976)

This volume has mainly been a collection of odds and ends, developing material that is related to Volume 9 (i.e., Part A), or my previous work. The unifying theme has been the role the geometry of manifolds plays in analytical mechanics (classical and quantum), control theory and (in the last chapter) economics.

I believe that there is an underlying unity to this material, despite the fact that it reaches out towards such diverse disciplines. (If A is the set of physicists, B the set of control engineers, C the set of economi$s\tau s$, let A # B denote the set of pairs (a,b) ε A × B that in some sense communicate with each other. In my experience,

$$(A \# B) \;=\; (A \# C) \;=\; (B \# C) \;=\; \text{(the empty set)}. \;)$$

Most of this material was written or thought about over a year ago. I probably would do it quite differently now. In all three areas, I did not write down all that I meant to-- which might mean a Part C! For example, in mechanics I want to understand the Lagrange-Liapounoff stability ideas and their relation to the symmetry-group theory ideas in a more

complete form than exists in the literature. (For example,
what is the general setting for the classical stability
properties of a top? Of a non-holonomic system such as a
bicycle?) I have recently understood the relation between
differential geometry and quantum mechanics in a more complete
form--what is involved is a sort of a higher order version of
"Poisson bracket" called the Moyal bracket. Geometrically,
this corresponds to what we geometers might call a "higher
order (in the sense of Ehresmann's jets) symplectic structure"
Similarly, the relations between "complete solutions of the
Hamilton-Jacobi equation", "canonical transformations
(classically) and unitary transformations (quantumly)", and
so forth, has to be thought out more systematically.

In my lectures this semester at Brown University I am
developing the Caratheodory approach to optimal control theory
in much more detail, and with a slightly different formalism,
closer to that used by the engineers, than in this volume.
This will probably appear as part of the Second Edition to
"Differential Geometry and the Calculus of Variations", which
is in preparation (and will be published by Math Sci Press
as part of this series, since Academic Press let it go out of
print and gave me the copyright).

The last chapters on Economics are obviously the most
incomplete. They were originally written as part of a volume
projected for this series, which was to have been called

"Optimization and Economic Theory on Manifolds". As I thought about this project, I realized that I was not satisfied enough with the standard material to commit it to an independent volume. (In fact, the book by Intriligator [1] already gives an excellent treatment.) I want to understand the relation between systems-control theory and economics in much more depth than has been done up to now. (Unfortunately, all the talk about applying control theory to economics has not produced very much. I am convinced that there is a strong relation, and one of my projects is to understand this better!) In the meantime, I have decided to publish the material in these chapters since it is not available elsewhere (at least from this point of view). The most distinctive topic is the algebraic treatment of the second order conditions for "constrained minima". After I wrote these chapters I realized there was a better organizing principle than the one I used--what the economists call the interpretation of "prices as Lagrange multipliers". Here is the idea in simpler form

Let X, Y be spaces, and let

$$\pi: X \to Y$$

be a map,

$$f: X \to R$$

be a real-valued function. Y is a subset of a real vector space V. (X is not necessarily finite dimensional; in the

calculus of variations, it will be a space of curves, surfaces,
etc.) Then, we construct two sorts of objects: Maps

$$S: Y \to R \quad ,$$

and

$$\gamma: Y \to X \quad ,$$

such that:

$S(y)$ = extremum value of the function $x \to f(x)$
<u>restricted to the fiber</u> $\pi^{-1}(x)$

$\gamma(y)$ is the value of x at which

$$x \to f(x,y)$$

has an extremum.

Much of "optimization theory" has to do with the <u>geometric</u>
and <u>differential calculus properties</u> of these two objects.
For example, "Lagrange multipliers" are values of $dS(y)$,
i.e., element of the dual space V^d. In economics they are
"prices". This interpretation of "Lagrange multipliers" as
"prices" is emphasized by Intriligator. In fact, one of the
nicest features of his book is that he describes very well
the <u>intuitive</u> content (from the point of view of a mathematical economist) of much of optimization theory, without the
precious and <u>too</u> precise and sophisticated language of the

Arrow-Debreu school. Thus, Intriligator's book is an excellent example of "Interdisciplinary Mathematics" from an
economist's point of view.

Bibliography

1.   R. Abraham and J. Marsden, Foundations of Mechanics,
     W.A. Benjamin, N.Y., 1966.

1.   B. Anderson and J. Moore, Linear optimal Control, Prentice-
     Hall, Englewood Cliffs, N.J., 1971.

1.   B. Anderson and S. Vongpanitlerd, Network Analysis and
     Synthesis, Prentice-Hall, Englewood Cliffs, N.J., 1973.

1.   V. Arnold, Sur la géométrie des groupes de Lie de
     dimension infinie et ses applications à l'hydrodynamique
     des fluides parfaits, Ann. Inst. Gren. 16, 319-361 (1966).

1.   A.M. Arthurs, Complementary Variational Principles,
     Oxford University Press, 1920.

1.   E. Artin, Geometric Algebra, Interscience Publishers, 1957.

1.   R. Brayton and J. Moser, Theory of Non-Linear Networks,
     Quart. App. Math. 22, 1-33, 81-104, 1964.

1.   R. Brockett, Finite Dimensional Linear Systems, Wiley,
     N.Y., 1965.

1.   P.F. Byrd and M.D. Friedman, Handbook of Elliptic Integrals,
     Springer-Verlag, Berlin, 1954.

1.   C. Caratheodory, Calculus of Variations and Partial
     Differential Equations of the First Order, Holden-Day,
     San Francisco, 1967.

1.   E. Cartan, Leçons sur les invariants intégraux, Hermann,
     Paris, 1922.

2.   E. Cartan, Sur les varietés à connexion affine et la
     théorie de la relativité generalisée, Annals. Ecole
     Normale, t.40 (1923), 325-412 (≡ Oeuvres, pt. III, vol. I).

1.   H. Corben and P. Stehle, Classical Mechanics, Wiley,
     N.Y., 1960.

1.   C. Desoer and E. Kuh, Basic Circuit Theory, McGraw-Hill,
     N.Y., 1969.

1.   E. Dierker, Topological Methods in Walrasian Economics,
     Springer-Verlag, Heidelberg, 1974.

481

1.   D. Ebin and J. Marsden, Groups of Diffeomorphisms and
      the Motion of an Incompressible Fluid, Ann. Math., 92,
      102-163 (1970).

1.   L. Fadeev and V. Zakharov, Korteweg-de Vries Equation:
      A Completely Integrable Hamiltonian System, Functional
      Anal. and App. 5, 1972, 280-287.

1.   H. Goldstein, Classical Mechanics, Addison-Wesley,
      Reading, Mass., 1951.

1.   W. Greub, Multilinear Algebra, Springer-Verlag, Berlin,
      1967.

1.   R. Hermann, Some Differential-Geometric Aspects of the
      Lagrange Variational Problem, Ill. J. Math., 6, 634-673
      (1962).

2.   R. Hermann, On the Accessibility Problem of Control
      Theory, Proceedings of Symposium on Diff. Eq., ed.,
      Lasalle and Lefschetz, Academic Press, N.Y., 1961.

3.   R. Hermann, Cartan Connections and the Equivalence
      Problem for Geometric Structures, in "Contributions to
      Differential Equations", 3 (1964), 199-248.

4.   R. Hermann, E. Cartan's Geometric Theory of Partial
      Differential Equations, Advances in Math., 1 (1965),
      265-317.

5.   R. Hermann, The Second Variational Formula for Variational
      Problems in Canonical Form, J. of Math and Mech., 16
      (1966), 473-492.

6.   R. Hermann, The Second Variation for Minimal Submanifolds,
      J. of Math. and Mech., 16 (1966(, 473-492.

7.   R. Hermann, Lie Groups for Physicists, W.A. Benjamin, Inc.,
      N.Y., 1966. (Abb: LGP).

8.   R. Hermann, Differential Geometry and the Calculus of
      Variations, Academic Press, N.Y., 1969. (Abb: DGCV).

9.   R. Hermann, Lie Algebras and Quantum Mechanics, W.A.
      Benjamin, N.Y., 1970. (Abb: LAQM).

10.   R. Hermann, Vector Bundles in Mathematical Physics,
      Parts I and II, W.A. Benjamin, N.Y., 1970. (Abb: VB).

11. R. Hermann, Lectures on Mathematical Physics, Vols. I
     and II, W.A. Benjamin, N.Y., 1970, 1972. (Abb: LMP).

12. R. Hermann, Geometric Ideas in Lie Group Harmonic Analysis
     Theory, Proceedings of the Washington Symposium on
     Symmetric Spaces, ed., W. Boothby and G. Weiss, Marcel
     Dekker, Inc., N.Y., 1972.

13. R. Hermann, "Spectrum-Generating Algebras in Classical
     Mechanics, I, II", J. Math. Phys. 13, 1972, 833-878.

14. R. Hermann, "Left Invariant, Geodesics and Classical
     Mechanics on Manifolds", J. Math. Phys. 13, 1972, 460.

15. R. Hermann, "Geometry, Physics and Systems", Marcel Dekker,
     N.Y., 1973. (Abb: GPS).

16. R. Hermann, "Physical Aspects of Lie Group Theory", Univ.
     of Montreal Press, Montreal, 1974. (Abb: PALG).

17. R. Hermann, Some Remarks on the Geometry of Systems,
     Proceedings of NATO Institute on Geometric and Algebraic
     Methods for Non-Linear Systems, Imperial College, D. Mayne,
     ed., Reidel Publishing Co. (Dordrecht, Holland and Boston,
     Mass.), 1973.

18. R. Hermann, Interdisciplinary Mathematics, Vols. 1-10,
     Math Sci Press, Brookline, Mass. (1973-75). (Abb: IM).

19. R. Hermann and N. Wallach, eds., Lie Groups: History,
     Frontiers and Applications, Vols. 1-4, Math Sci Press,
     Brookline, Mass. (1975-76). (Abb: LG).

20. R. Hermann, Geodesics of Singular Riemannian Metrics,
     Bull. AMS, 79 (1973), 780-782.

 1. M. Hirsch and S. Smale, Differential Equations, Academic
     Press, N.Y., 1974.

 1. M. Intriligator, Mathematical Optimization and Economic
     Theory, Prentice Hall, Englewood Cliffs, N.J., 1971.

 1. J. Keller, Corrected Bohr-Sommerfeld Quantum Conditions
     for Non-Separable Systems, Ann. Phys. 4 (1958), 180-188.

 1. B. Kostant, Quantization and Unitary Representations,
     Lectures in Math., Vol. 170, Springer-Verlag, Heidelberg,
     1920.

1.   H.W. Kuhn and A.W. Tucker, "Nonlinear Programming", in
     Proceedings of the Second Berkeley Symposium on Mathe-
     matical Statistics and Probability, ed., J. Negman,
     Berkeley, Calif., Univ. of California Press, 1951.

1.   C. Lanczos, The Variational Principles of Mechanics,
     Univ. of Toronto Press, 1954.

1.   S. Lang, Algebra, Addison-Wesley, Reading, Mass., 1967.

1.   E.B. Lee and F. Markus, Foundations of Optimal Control
     Theory, J. Wiley, N.Y., 1967.

1.   V. Maslow, Theorie des Perturbations et Methods Asymp-
     totiques, Dunod, Paris, 1972.

1.   J. Milnor, Morse Theory, Princeton Univ. Press, Princeton,
     N.J., 1963.

1.   M. Morse, Calculus of Variations in the Large, Am. Math.
     Soc., Providence, R.I., 1935.

1.   G. Oster and A. Perelson, Chemical Reaction Dynamics,
     Arch. Rat. Mech. Anal., 55 (1974), 230-274.

1.   L. Pars, A Treatise on Analytical Dynamics, Wiley, N.Y.,
     1965.

1.   H. Poincaré, New Methods of Celestial Mechanics, National
     Aeronautics and Space Administration, Washington, D.C.,
     1967.

1.   R. Prosser, J. Math. Phys, 5 (1964), 701.

1.   P.A. Samuelson, Foundations of Economic Analysis, Harvard
     University Press, Cambridge, Mass. (1947).

1.   I.E. Segal, Mathematical Problems of Relativistic Physics,
     Am. Math. Soc., Providence, R.I., 1963.

1.   A. Sommerfeld, Mechanics, Academic Press, N.Y., 1950.

1.   J.M. Souriau, Structure des systemes dynamiques, Dunod,
     Paris, 1970.

1.   L. Van Hove, Sur certaines representations unitaires d'un
     groupe infinis de transformations, Mem. Acad. Roy. Belg.
     26 (1951).

1.   E. Whittaker, Analytical Dynamics, Cambridge Univ. Press,
     London, 1959.

8480-78-2
5-32